'This is a very rare book. A scientific book about climate change that deals both with the science and with our own personal response to this science. It does all this supremely well, and should be compulsory reading for both sceptics and believers. However, it does so much more; it is a book of great modesty and humanity. It uses climate change to ask broader questions about our own beliefs, assumptions and prejudices, and how we make individual and collective decisions.'

Chris Mottershead, *Distinguished Advisor, BP plc*

'In this personal and deeply reflective book, a distinguished climate researcher shows why it may be both wrong and frustrating to keep asking what we can do for climate change. Exploring the many meanings of climate in culture, Hulme asks instead what climate change can do for us. Uncertainty and ambiguity emerge here as resources, because they force us to confront those things we really want – not safety in some distant, contested future, but justice and self-understanding now. Without downplaying its seriousness, Hulme demotes climate change from ultimate threat to constant companion, whose murmurs unlock in us the instinct for justice and equality.'

Sheila Jasanoff, Pforzheimer Professor of Science and Technology Studies, Harvard University

'This book is a "must read" for anyone interested in the relationship between science and society. As we know from the controversies over GM crops and MMR, by the time science hits the headlines, and therefore the public consciousness, it's always about much more than the science. This fascinating book shines a light on this process by revealing how climate change has been transformed from a physical phenomenon, measurable and observable by scientists, into a social, cultural and political one.

Everyone must surely recognise Hulme's description of the way that climate change has become a kind of Christmas tree onto which we all hang our favourite baubles, and Hulme highlights the way in which the issue has been appropriated by so many different groups to promote their own causes. Believers in turning the clock forward and using more advanced technology, and those who argue that we should turn the clock back and live more simply, can equally claim that climate change supports their case.

Over the past few years Hulme has bravely spoken out against what some have described as "climate porn": the tendency of some sections of the scientific community and the media to present climate change in ever more catastrophic and apocalyptic terms. This book elaborates on Hulme's hostility to the language of "imminent peril" and calls for a different discourse.

This book is so important because Mike Hulme cannot be dismissed as a sceptic, yet he is calling for a complete change in the way we discuss climate change. Whether or not people agree with his conclusions – this book is a challenging, thought-provoking and radical way to kick-start that discussion.'

Fiona Fox, Director, Science Media Centre, London

'With empirical experience that includes seven years leading the influential Tyndall Centre, Professor Hulme here argues that science alone is insufficient to face climate change. We also "need to reveal the creative psychological, spiritual and ethical work that climate change can do and is doing for us". It is the very "intractability of climate change", its sociological status as a 'wicked' problematique, that requires us to reappraise the "myths" or foundational belief systems in which the science unfolds. That returns Hulme to the bottom-line question: "What is the human project ultimately about?" – and herein resides this book's distinctive importance.'

Alastair McIntosh, author of 'Hell and High Water:
Climate change, hope and the human condition', and
Visiting Professor of Human Ecology at the Department
of Geography and Sociology, University of Strathclyde

'A much needed re-examination of the idea of climate change from a vantage point that takes its cultural co-ordinates as seriously as its physical properties. Through the twin lenses of scientific scrutiny and rhetorical analysis, Mike Hulme helps us to see just why we disagree about climate change and what we can do about it. With wisdom, wit and winsome writing, he shows us that debates about climate change turn out to be disputes about ourselves – our hopes, our fears, our aspirations, our identity. Hindsight, insight and foresight combine to make this book a rare treat.'

David N. Livingstone, Professor of Historical Geography, Queen's
University Belfast

'In the crowded and noisy world of climate change publications, this will stand tall. Mike Hulme speaks with the calm yet authoritative voice of the integrationist. He sees climate change as both a scientific and a moral issue, challenging our presumed right to be "human" to our offspring and to the pulsating web of life that sustains habitability for all living beings. As a unique species, we have the power to create intolerable conditions for the majority of our descendents. Yet we also have the scientific knowledge, the economic strength, and the political capacity to change direction and put a stop to avoidable calamity. This readable book provides us with the necessary argument and strategy to follow the latter course.'

Tim O'Riordan, Emeritus Professor of Environmental Sciences,
University of East Anglia

WHY WE DISAGREE ABOUT CLIMATE CHANGE

Climate change is not 'a problem' waiting for 'a solution'. It is an environmental, cultural and political phenomenon which is reshaping the way we think about ourselves, our societies and humanity's place on Earth. Drawing upon twenty-five years of professional work as an international climate change scientist and public commentator, Mike Hulme provides a unique insider's account of the emergence of this phenomenon and the diverse ways in which it is understood. He uses different standpoints from science, economics, faith, psychology, communication, sociology, politics and development to explain why we disagree about climate change. In this way he shows that climate change, far from being simply an 'issue' or a 'threat', can act as a catalyst to revise our perception of our place in the world. *Why We Disagree About Climate Change* is an important contribution to the ongoing debate over climate change and its likely impact on our lives.

MIKE HULME is Professor of Climate Change in the School of Environmental Sciences at the University of East Anglia (UEA), and founding Director of the Tyndall Centre for Climate Change Research. He has published over a hundred peer-reviewed journal papers and over thirty books or book chapters on climate change topics. He has prepared climate scenarios and reports for the UK Government, the European Commission, UNEP, UNDP, WWF International and the IPCC. He is currently leading the EU integrated research project ADAM (Adaptation and Mitigation Strategies), a consortium of twenty-six institutes contributing research to the development of EU climate policy during the period 2006–9. He co-edits the journal *Global Environmental Change* and is Editor-in-Chief of Wiley's *Interdisciplinary Reviews: Climate Change*. He is a frequent speaker about climate change at academic, professional and public events, and writes frequently for the media.

Why We Disagree About Climate Change

UNDERSTANDING CONTROVERSY, INACTION AND OPPORTUNITY

Mike Hulme

School of Environmental Sciences,
University of East Anglia

CAMBRIDGE
UNIVERSITY PRESS

CAMBRIDGE UNIVERSITY PRESS

Cambridge, New York, Melbourne, Madrid, Cape Town, Singapore, São Paulo, Delhi

Cambridge University Press
The Edinburgh Building, Cambridge CB2 8RU, UK

Published in the United States of America by Cambridge University Press, New York

www.cambridge.org
Information on this title: www.cambridge.org/9780521727327

First published 2009

Printed in the United Kingdom at the University Press, Cambridge

A catalogue record for this publication is available from the British Library

Library of Congress Cataloguing in Publication data
Hulme, M. (Michael)
Why we disagree about climate change : understanding controversy,
inaction and opportunity / Mike Hulme.
p.   cm.
Includes bibliographical references and index.
ISBN 978-0-521-89869-0 (hbk.)
1. Climatic changes. 2. Climatic changes–Government policy.
3. Environmental policy. I. Title.
QC981.8.C5H825   2009
363.738'74–dc22
2009004684

ISBN 978-0-521-89869-0 hardback
ISBN 978-0-521-72732-7 paperback

To my father, Ralph Hulme (1924–1989),
who taught me that disagreeing was a form of learning

A good place to look for wisdom … is where you least expect to find it: in the minds of your opponents.

Jonathan Haidt (2006)

Contents

List of Figures

List of Tables

List of Boxes

Acknowledgments

This book was conceived in February 2003 while I was browsing in the bookshop of Church House in London, was subsequently gestated during my remaining tenure as Director of the Tyndall Centre, and was brought to fruition during a period of study leave granted me by the University of East Anglia (UEA). I would like to thank all of my colleagues in the Tyndall Centre for introducing me to many of the ideas discussed here and for providing such a stimulating environment in which to observe, think and talk about climate change. Many of these colleagues have discussed with me some of the ideas contained in this book – and at times very helpfully disagreed! I would like to thank in particular those who have read and commented on draft extracts or chapters: Tim O'Riordan, Irene Lorenzoni, Natasha Grist, John Turnpenny, Sam Randalls, Tom Lowe, Asher Minns, Saffron O'Neill, Lorraine Whitmarsh, Jacqueline de Chazal, Nick Brooks, David Livingstone, Neil Adger, Joe Smith, Dave Ockwell, Christopher Shaw, Chuks Okereke, Sarah Dry, Neil Jennings, Don Nelson and Mark Charlesworth. Suraje Dessai is deserving of particular appreciation, not only for commenting on drafts of individual chapters, but more importantly for the countless engaging and provocative conversations about climate change that we have enjoyed over nine fruitful years of working together. The book also benefited from invitations received from Science and Technology Policy Research at the University of Sussex, the Faraday Institute at the University of Cambridge, and the Lincoln Theological Institute at the University of Manchester, where I was able to present some of the ideas contained here.

My – as yet unfinished – graduate diploma in the UEA School of History opened up for me new perspectives on the nature and writing of history, perspectives that were essential in equipping me for the task of writing this book. I also thank the 2007/8 cohort of forty-four Masters students at UEA who took my climate change module while this book was taking shape. A number of my ideas were tried out on these students and the seminar debates that we enjoyed offered some good new perspectives on why we disagree about climate change. Phil Judge did a thoroughly professional job in re-drawing a number of the figures in the book, while Chris Harrison at Cambridge University Press embraced the project with enthusiasm and, together with Philip Good, guided it through its commissioning and production stages. Finally I thank Gill and Emma for being, as always, my most honest critics and my most loyal supporters.

Foreword

When I first entered the field of climate change policy research, a little over two decades ago, I was warned by a former deputy administrator of the US Environmental Protection Agency that I was wasting my time because: 'Climate change will never be a major public policy issue.' He advanced three reasons for this: 'The science is too uncertain, the impacts are too far in the future, and there is no readily identifiable villain.' My response was that these were exactly the kinds of reasons why climate change would become a major policy issue. It was precisely the plasticity of climate change – its ability to be many things to many people – that would ensure its claim to sustained public attention.

Ten years later, hard on the heels of the Kyoto Protocol, I led the publication of a state-of-the-art report on the social science research relevant to climate change[1] that confirmed what we have subsequently recognised as the 'wickedness' of climate change as an issue. Wickedness in this sense is not a moral judgement (although to some people climate change is the consequence of an unethical industrial

[1] Rayner, S. and Malone, E.L. (eds) (1998) *Human choice and climate change* (4 vols). Battelle Press: Columbus, OH.

lifestyle). Originating in the study of urban planning,[2] it is a way of describing problems of mind-bending complexity, characterised by 'contradictory certitudes' and thus defying elegant, consensual solutions.

As Mike Hulme lays out in this volume, a further decade on, climate change is not so much a discrete problem to be solved as it is a condition under which human beings will have to make choices about such matters as priorities for economic development and the way we govern ourselves. Our recognition of climate change as a threat to the ways of life to which we are accustomed and which we value depends on our views of Nature, our judgements about scientific analysis, our perceptions of risk, and our ideas about what is at stake – economic growth, national sovereignty, species extinction, or the lives of poor people in marginal environments of developing countries – and whether it is ethically, politically, or economically justifiable to make trade-offs between these.

Even when scientists, politicians and publics agree on the basic principles and most robust findings of climate science, there is still plenty of room for disagreement about what the implications of that science are for action. The notion that science 'drives' consensus on policy and that better science will settle our differences ignores the roots of these differences in political, national, organisational, religious and intellectual culture. What is taken for granted among one group of people is 'uncomfortable knowledge' that is hard for another to accept because of its implications for ideas and resource commitments that they hold dear.[3]

Why We Disagree About Climate Change advances this sociological perspective on climate debates in at least two important ways. Firstly, it is written by a distinguished climate scientist. In the field of

---

[2] Rittel, H. and Webber, M. (1973) Dilemmas in a general theory of planning. *Policy Sciences* 4(2), 155–69.

[3] Thompson, M. and Rayner, S. (1998) Cultural discourses, pp. 265–344, in Rayner and Malone (eds), *Human choice and climate change.*

climate change, scientists, politicians and journalists are likely to take the views of natural scientists more seriously than those of anthropologists, sociologists or political scientists, even when the issues are concerned with the behaviour of social systems rather than natural systems. This applies even when those scientists' own specialisations have only a tangential connection with climate science (as is the case with at least two former Chief Scientists of the UK Government who have insisted that climate modelling is robust beyond question). As science has become the ultimate source of legitimacy in contemporary society, many of its practitioners act as though they are guardians of unquestionable doctrine, telling us that 'the science of climate change is settled' (and, by implication, climate policy is settled also). On the other hand, climate science specialists, like Hulme, who are closer to the point of production of climatological knowledge, are fully aware of the inevitable gaps and ambiguities in the science, and therefore of the necessity for strategic social and political dialogue about how to respond in the light of epistemological uncertainty and competing social values.

But secondly, and perhaps more importantly, Mike Hulme has done more than simply lend his scientific authority to the sociological and anthropological insights into climate debates. He goes further than even most social scientists have dared, by pointing out that climate change debates are more than merely a peg on which different interest groups can hang their particular agendas. Rather, climate change provides a much needed arena and stimulus for public discussion of the big issues of our time.

Climate is more than just a coercive resource to be mobilised behind different visions of humanity and its future. It has become the key narrative within which political issues from the local to the global are framed. In that sense, debate around climate has succeeded debate around capital and social class as the organising theme of political discourse in contemporary society. Hulme concludes that: 'Rather than placing ourselves in a "fight against climate change" we need a more constructive and imaginative engagement with the idea of climate change.' Paraphrasing John F. Kennedy, he argues that 'we need to ask not what we can do for climate change, but to ask what climate change can do for us?' We should, he says, 'use climate change both as a magnifying glass and as a mirror': as a magnifier, to focus our attention on the long-term implications of short-term choices in the context of material realities and social values; and as a mirror, to 'attend more closely to what we really want to achieve for ourselves and for humanity'.

Hence, the plasticity of what is conveyed by the term 'climate change' is not merely an obstacle to agreement on elegant solutions (such as carbon trading), or even to the emergence of what have been variously called 'clumsy solutions' or 'incompletely theorised agreements'.[4] It provides the public arena and the vocabulary, even (perhaps ironically in view of the dangers presented by climate itself) the 'safe' virtual space in which people can confront each other with rival world-views, competing ideals of the social good, and conflicting economic commitments. As Hulme clearly demonstrates in this book, reducing this rich and complex public discourse about the nature of 'the good' simply to a technical or even political debate about the 'acceptable' level of carbon dioxide in the atmosphere is simply to miss the point.

Steve Rayner
James Martin Professor of Science and Civilization,
University of Oxford

---

[4] Verweij, M. and Thompson, M. (eds) (2006) *Clumsy solutions for a complex world: governance, politics and plural perceptions.* Palgrave Macmillan: Basingstoke, UK; and Sunstein, C. (1995) Incompletely theorized agreements. *Harvard Law Review* 108, 1733–72.

Preface

Why We Disagree About Climate Change is a book about the idea of climate change: where it came from, what it means to different people in different places, and why we disagree about it. It is a book which also develops a different way of approaching the idea of climate change and of working with it.

I deliberately present climate change as an idea as much as I treat it as a physical phenomenon that can be observed, quantified and measured. This latter framing is how climate change is mostly understood by scientists, and how science has presented climate change to society over recent decades. But, as society has been increasingly confronted with the observable realities of climate change and heard of the dangers that scientists claim lie ahead, climate change has moved from being predominantly a physical phenomenon to being simultaneously a social phenomenon. And these two phenomena are very different. As we have slowly, and at times reluctantly, realised that humanity has become an active agent in the reshaping of physical climates around the world, so our cultural, social, political and ethical practices are reinterpreting what climate change means. Far from simply being a change in physical climates – a change in the sequences of weather experienced in given places – climate change has become an idea that now travels well beyond its origins in the natural sciences. And as this idea meets new cultures on its travels and encounters the worlds of politics, economics, popular culture, commerce and religion – often through the interposing role of the media – climate change takes on new meanings and serves new purposes.

In *Why We Disagree About Climate Change*, I examine this mutating idea of climate change. I do so using the concepts, tools and languages of the sciences, social sciences and humanities, and the discourses and practices of economics, politics and religion. As we examine climate change from these different vantage points, we begin to see that – depending on who one is and where one stands – the idea of climate change carries quite different meanings and seems to imply quite different courses of action. These differences of perspective are rooted much more deeply than (merely) in contrasting interpretations of the scientific narrative of climate change. Our discordant conversations about climate change reveal, at a deeper level, all that makes for diversity, creativity and conflict within the human story – our different attitudes to risk, technology and well-being; our different ethical, ideological and political beliefs; our different interpretations of the past and our competing visions of the future. This discord, in the context of climate change, has been described by the novelist Ian McEwan: 'Can we agree among ourselves? We are a clever but quarrelsome species – in our public discourses we can sound like a rookery in full throat.'[1] If we are to understand climate change and if we are to use climate change constructively in our politics, we must first hear and understand these discordant voices, these multifarious human beliefs, values, attitudes, aspirations and behaviours. And, especially, we must understand what climate change signifies for these important dimensions of human living and human character.

---

[1] p. 3 in British Council (2005) *Talking about climate change*. British Council: Manchester, UK.

To illustrate what I mean, let me cite four contemporary and contrasting ways of narrating the significance of climate change – just some of the more salient discourses currently in circulation.

Climate change as a battleground between different philosophies and practices of science, between different ways of knowing. 'Climate change as scientific controversy' is a compelling discourse to which the media and other social actors are readily attracted. Although the controversy is allegedly about science, very often scientific disputes about climate change end up being used as a proxy for much deeper conflicts between alternative visions of the future and competing centres of authority in society.

Climate change as justification for the commodification of the atmosphere and, especially, for the commodification of the gas, carbon dioxide. In this frame, climate change is viewed as the latest rationale for converting a public commons into a privatised asset – in this case, the global atmosphere. 'Ownership rights' to emit carbon dioxide are allocated or auctioned between entities, alongside the attendant machinery of the market which prices and regulates the commodity.

Climate change as the inspiration for a global network of new, or reinvigorated, social movements. Seeing climate change as a manifestation of the nefarious practices of globalisation, this framing warrants the emergence of new forms of activism, both elite and popular, to challenge these practices and to catalyse change in political, social and economic behaviour.

Climate change as a threat to ethnic, national and global security. The rhetoric associated with this framing compares climate change (unfavourably) with the threats posed by international terrorism, warranting a new form of geo-diplomacy at the highest levels of government. This framing has been espoused especially by the UK Government in recent years, and led in April 2007 to the first debate about climate change to be held at the United Nations Security Council.

These examples of sites of scientific, economic, social and political conflict and innovation illustrate that the idea of climate change possesses a certain plasticity – or at least that the idea of climate change has been constructed in such a way as to ensure that it possesses this quality of plasticity. Such an attribute allows climate change easily to be appropriated in support of a wide range of ideological projects. Climate change can be framed, can be moulded, in many different ways. Sometimes these frames complement each other, yet often they appear to conflict.

All of the above suggests that, far from starting with ignorance and ending with certainty, the story of climate change is much more interesting. It is a story about the meeting of Nature and Culture,[2] about how humans are central actors in both of these realms, and about how we are continually creating and re-creating both Nature and Culture. Climate change is not simply a 'fact' waiting to be discovered, proved or disproved using the tenets and methods of science. Neither is climate change a problem waiting for a solution, any more than the clashes of political ideologies or the disputes between religious beliefs are problems waiting to be solved.

The full story of climate change is the unfolding story of an idea and how this idea is changing the way we think, feel and act. Not only is climate change altering our physical world, but the idea of climate change is altering our social worlds. And this idea is reaching farther and farther across these social worlds.[3] Rather than asking 'How do we solve climate change?' we need to turn the question around and ask 'How does the idea of climate change alter the way we arrive at and achieve our personal aspirations and our collective social goals?' By understanding why we disagree about climate change we will also

[2] I adopt capitalisation for the nouns 'Nature' and 'Culture', here and elsewhere in the book, to signify that I am treating them as unique entities rather than as a class of entities. Although there are many cultures, the idea of Culture is singular. Similarly, while we may recognise many different natures around us, the constructed idea of Nature is singular.

[3] A poll of 22,000 citizens across twenty-one countries conducted for the BBC World Service in September 2007 revealed that nearly 80 per cent of those surveyed believe that human activities are a significant cause of climate change.

better understand what it takes to live sustainably on a crowded finite planet inhabited by a 'quarrelsome species'.

The account of climate change that I present in this book emerges from my own encounter with climate change over the last thirty years. This encounter started while I was a university student, continued during my time as a post-doctoral researcher and, more recently, has persisted through my roles as a professor, research leader, educator and public speaker. To understand why and how I have written this book it is necessary to understand the different stages of this encounter. I briefly relate my evolving relationship with climate change through identifying six stages in my encounter. These personal and professional experiences have shaped the way I now view climate change. This journey is also worth relating because the period through which I have travelled – from the late 1970s to today – coincides with the transformation of climate change from an object of largely scientific professional interest into a topic of daily and worldwide popular discourse.

I first encountered climate change when I was a geography student in England in the late 1970s. The idea of climate change seemed to embody my twin enthusiasms for geography and history, and the first full-length book I read on the subject was Hubert Lamb's 1977 *Climate Change: Present, past and future*. This started me on the first stage of my journey, covering the period from 1978 to 1988; a stage which I describe as '*Youthful Idealism*'. My undergraduate geography degree at the University of Durham introduced me to the idea that climates change on human time-scales. I well remember being taught at that time about claims of an approaching and seemingly imminent Ice Age, which struck me then, as they still do today, as being far-fetched.

It was the subsequent opportunity to pursue a PhD at the University of Wales, researching recent rainfall changes in Sudan, which convinced me that studying climate change could be a way of satisfying my humanitarian ideals, born out of my Christian beliefs.[4] In Sudan I witnessed, at first hand, the suffering and devastation caused by the Sahelian drought of the early 1980s, and came to believe that a better understanding of climate variability and change could help alleviate human suffering. Climate change also provided me with the gateway into my first professional appointment, as a lecturer in geography at the University of Salford in 1984. This was to be the stepping stone to me later securing a post-doctoral research position under the inspiring Professor Tom Wigley at the Climatic Research Unit in the School of Environmental Sciences at the University of East Anglia in Norwich, UK.

For my PhD, and then later at Norwich, I became immersed in studying climate change through the application of quantitative, especially statistical, methods. This was the second stage in my journey with climate change; a stage I describe as '*Quantitative Analysis*' (*c.* 1981–98). This period saw my major analytical contribution to the science of climate change. I worked with worldwide observational climate data to examine global and regional trends in precipitation and with the small, but growing, international network of climate modelling centres – especially the Hadley Centre in the UK – to evaluate the performance of a new generation of climate models. In this mode of working and publishing, I saw climate change as a physical phenomenon which could be revealed and defined by quantitative data and understood and predicted by models.

Embedded within this analytical period was another important stage in my journey, in which I came to see climate change in terms of '*Political Ideology*' (*c.* 1984–90). I came to view global climate change caused by greenhouse gas emissions as a manifestation of a free-market, consumption-driven, capitalist economy – an ideology to

---

[4] I have written elsewhere about this aspect of my story. See Hulme, M. (2009) A belief in climate. Chapter in Berry, S. (ed.), *Real scientists, real faith*. Lion Hudson: London.

which I was opposed. I recollect now that this opposition was an explicit ideological frame I used when teaching my course on contemporary climate change to final-year undergraduate geography students at the University of Salford between 1985 and 1988. This way of relating to climate change was a formative influence on (or reflection of) my political thinking during the decade of Thatcherite conservatism in the UK. I subsequently joined the British Labour Party in 1990.

My intellectual and emotional relationship with climate change also shaped me in other ways. After climate change first burst into public consciousness during the 'greenhouse summer' of 1988, I began to examine my own contribution to the causes of climate change, marking the fourth stage of my journey – '*Lifestyle Choices*'. From 1988 onwards, I was motivated – and still am – to make personal commitments to reduce my carbon and ecological footprints; commitments which influenced my behaviour with regard to energy demand, house purchases and modes of transportation. For example, in 2000 I purchased one of the first-generation Toyota Prius hybrids as our family car. During this period, in addition to my professional speaking, I also started giving talks to local Christian and community groups about the causes and implications of climate change.

Through a series of research contracts with the European Union, the UK Government and other international funding organisations, a fifth stage in my journey with climate change opened up during the 1990s: a stage which I call '*Scenarios for Policy*' (c. 1993–2002). These contracts required me to lead the development of a series of climate change scenarios for the UK, for Europe, for the Intergovernmental Panel on Climate Change, and for other countries and regions in Africa and Asia. In particular, under contract to the UK Department of the Environment, I led the design and delivery of the national UK climate change scenarios in 1998 and again in 2002. These scenario activities drew upon my analytical expertise with observed data and my familiarity with climate models, and took my professional career in a direction closer to policy assessments and societal decision making.

During the important international climate policy-shaping years of the late 1990s and early 2000s, I began to see climate change increasingly as an issue of public policy and strategic decision making and less as an object of detached quantitative scientific analysis.

The sixth, and last, stage in my journey started in 1999, when I was asked to lead a consortium of UK universities bidding for a new national climate change research centre; a centre which I subsequently named after the nineteenth-century Irish scientist John Tyndall. The Tyndall Centre was designed and operated as an interdisciplinary enterprise drawing upon the natural and social sciences, economics and engineering to undertake research into the ways in which societies around the world might respond to changing climate. I began to see the bigger picture of how climate change had been initially constructed as an environmental science 'problem', but how this idea of climate change was now increasingly interpreted and reinterpreted in different ways by different social actors.

I gave public talks about climate change to many of these different organisations and interest groups – local authorities, business associations, citizens' groups, government civil servants – and offered testimonies to parliamentary hearings. I gave talks and participated in workshops around the world, in countries such as the Cayman Islands, China, Cyprus, the Czech Republic, Germany, India, Latvia, Moldova, the USA and Zimbabwe. I began to see that climate change meant very different things to different people, depending on their political, social and cultural settings. As the public discourses surrounding climate change multiplied and diversified after 2005 in the UK and elsewhere – I believe that the British Government's conference on Avoiding Dangerous Climate Change held in Exeter in February 2005 was influential in this regard – I became dissatisfied with the earlier ways in which I had related to climate change. I became fascinated by the malleability of the idea of climate change as it came to be appropriated in support of so many different causes.

This final stage of my journey – a stage I call '*Cultural Enlightenment*' – crystallised through studying for a part-time graduate diploma in history at my university. This introduced me to a new range of intellectual traditions: in environmental, imperial and Enlightenment history, in the history of science, and in the sociology of scientific knowledge. It is on the foundations of these historical readings and my insights gained through leading the Tyndall Centre over seven years, combined with the earlier stages of my research career stretching back to the early 1980s, that this book – *Why We Disagree About Climate Change* – is built.

Since I have narrated my personal journey with the idea of climate change, I should also state clearly my own position with regard to the phenomenon, in case I am misunderstood. I believe that the risks posed to people and places by the physical attributes of climate are tangible, are serious, and require constantly improving forms of human intervention and management. I believe that the physical functions of global climate and, consequently, the parameters of local weather are changing (largely) under the influence of the changing composition of the atmosphere caused by an array of human activities. And I believe that changes in climatic risks induced by such global climate change are also important and serious. We do well to minimise these risks by reducing the vulnerability of those exposed to them and by minimising further changes to the composition of the world's atmosphere. Yet I do not believe that the way we have framed these goals – most significantly through the UN Framework Convention on Climate Change and the Kyoto Protocol – is the only way of doing so. Nor do I believe it is necessarily the most appropriate way. I feel uncomfortable that climate change is widely reported through the language of catastrophe and imminent peril, as 'the greatest problem facing humanity', which seeks to trump all others. I believe that such reporting both detracts from what science is good at revealing to us and diminishes the many other ways of thinking, feeling and knowing about climate which are also essential elements in personal and collective decision making.

Why We Disagree About Climate Change is my attempt to articulate the reasons for these beliefs and to re-situate the idea of climate change more honestly as the subject of a more creative and less pejorative discourse. On the other hand, it is an account that possesses all the limitations of a book written by one person about ideas which appear differently depending on context and perspective. In my attempts to understand why people around the world disagree about climate change I am limited by my own position – as an Englishman trained as a geographer, employed as a university professor, holding orthodox Christian beliefs, exercising democratic socialist political preferences and living in a European country. Despite a career-long exposure to different people's understandings, descriptions and experiences of climate change that I described above, I cannot escape the biases of my position. Readers are invited both to recognise these biases – most notably the prominence given to examples and literature from the UK – but also to look beyond them, to the underlying arguments being made, and to re-examine these arguments according to their own contexts and beliefs. And to recognise that disagreeing is a form of learning.

Why We Disagree About Climate Change consists of ten chapters. The first two chapters set the scene for the rest of the book. Chapter 1 – *The Social Meanings of Climate* – explores the different ways in which societies, over time, have constructed the idea of climate, and how they have related to its physical attributes. It offers what I might call a brief cultural reading of the history of climate. Some ideological purposes to which the idea of climate has previously been put are suggested. Chapter 1 also touches on the different ways in which changes in climate have been invoked to tell the story of the rise and fall of human civilisations. Chapter 2 – *The Discovery of Climate Change* – explains how we have come to believe that humans are an active agent in changing the physical properties of climate; what I might call a journey of scientific discovery. I establish the scientific genealogy of the idea that climates change over time and that human actions aggregated across the world can affect the functioning of the climate system. It contrasts the earlier nineteenth-century reading of climate change as purely natural and occurring over geological time-scales with the more recent appreciation that climates change on the time-scales of human generations and can partly be moulded by human actions.

The following seven chapters then examine the idea of climate change from seven different standpoints, offering in each case reasons why we disagree about climate change. The seven facets, or lenses, through which I approach the subject are, respectively: science, economics, religion, psychology, media, development and governance. Chapter 3 – *The Performance of Science* – claims that one of the reasons we disagree about climate change is because we understand science and scientific knowledge in different ways. Science thrives on disagreement; indeed science can only progress through disagreement and challenge. But disagreements presented as disputes about scientific evidence, theory or prediction may often be rooted in more fundamental differences between the protagonists. These may be differences about epistemology, about values, or about the role of science in policy making. This chapter examines the changing nature of science and what significance this has for disagreements within scientific discourse about the existence, causes and consequences of human-induced climate change.

Chapter 4 – *The Endowment of Value* – turns the attention to economics, and argues that one of the reasons we disagree about climate change is because we value things differently. There are many different ways in which individuals and societies ascribe value to activities, people, assets and resources. This chapter explores some of these different frameworks and explains why the choice of valuation framework is so important when deciding what to do about climate change. Quite radically different prognoses for addressing climate change emerge, depending on what value system is adopted. It is only a short step from considering what things we value and why we value them, to understanding the role of ethics, spirituality and theology in climate change debates. Chapter 5 – *The Things We Believe* – suggests that one of the reasons we disagree about climate change is because we believe different things about ourselves, the universe and our place in the universe. These beliefs have a profound influence on our attitudes, our behaviour and our policies. This chapter explores the different ways in which the major world religions have engaged with climate change, and also how other large-scale collective movements have recognised spiritual or non-material dimensions of the phenomenon.

We also disagree about climate change because we worry about different things; an idea explored in Chapter 6 – *The Things We Fear.* There is a long history of humans relating to climate in pathological terms. Climate-related risks continue to surprise and shock us. A prospective, and not fully predictable, change in climate therefore offers fertile territory for the heightening of these fears. This chapter examines the construction of risk around climate change, drawing upon insights from social and behavioural psychology, risk perception and cultural theory. Chapter 7 – *The Communication of Risk* – considers the ways in which knowledge is communicated and shaped by the media, and argues that one of the reasons we disagree about climate change is because we receive multiple and conflicting messages about climate change and interpret them in different ways. The public and policy discourses of climate change are heavily influenced by the way that scientific knowledge is framed, by the language of climate change risks, and by the vested interests of communication actors. This chapter examines the ways in which climate change has been represented in the media, by campaigning organisations, and by advertisers. It contrasts these representations with the construction and identification of climate change by the formal knowledge community.

Our views of welfare and development and how we think about progress are powerful shapers of attitudes to climate change.

Chapter 8 – *The Challenges of Development* – suggests that one of the reasons we disagree about climate change is because we prioritise development goals differently. Our definition of poverty and how we understand the inequalities in our world are also important. This chapter outlines some of these different views and approaches and explains why an understanding of climate change cannot be separated from an understanding of development. We also disagree about climate change because our political ideologies and our views of appropriate forms of governance are different; the subject of Chapter 9 – *The Way We Govern*. Climate change has been a live public policy issue since the late 1980s. The Kyoto Protocol, negotiated in 1997, has been the benchmark international agreement for shaping the goals of (and disputes around) mitigation policies. The last few years have seen a growing attention to whether or not a new genre of adaptation policies is necessary or desirable. This chapter explores the various ways in which governments have approached the design and implementation of climate policy – at local, national and international scales – and at the rise of non-state actors in climate governance.

The final chapter of the book, Chapter 10 – *Beyond Climate Change* – offers a perspective on climate change which transcends the categories and disagreements explored in earlier chapters. It looks beyond climate change. The chapter argues that climate change is not a problem that can be solved in the sense that, for example, technical and political resources were mobilised to 'solve' the problem of stratospheric ozone depletion. We need to approach the idea of climate change from a different vantage point. We need to reveal the creative psychological, spiritual and ethical work that climate change can do and is doing for us. By understanding the ways in which climate change connects with these foundational human attributes we open up a way of re-situating culture and the human spirit at the heart of our understanding of climate. Human beings are more than material objects, and climate is more than a physical entity. Rather than catalysing disagreements about how, when and where to tackle climate change, the idea of climate change is an imaginative resource around which our collective and personal identities and projects can, and should, take shape.

Finally, a word about the terms used in the book to describe the phenomenon of climate change. The three most commonly used shorthand descriptions are 'climate change',[5] 'global warming' and the 'greenhouse effect'. My colleague at the University of East Anglia, Lorraine Whitmarsh, has explored how and why these terms signify different things among members of the British public.[6] Each of these terms has a technical meaning (in the case of 'climate change' two different technical meanings!)[7] which may differ from the implied or affective popular meaning. And as these terms get translated differently across linguistic and cultural barriers they take on further nuances of meaning different from the formal definitions of scientific English.

I use the term 'climate change'[8] to mean a past, present or future change in climate, with the implication that the predominant – but

[5] In recent years it has become increasingly common in popular discourse to preface the term 'climate change' with the adjective 'catastrophic'. This almost suggests a fourth generic description of the phenomenon. Chapter 7: *The Communication of Risk* examines some of the reasons for this new linguistic compound and its implications.

[6] Whitmarsh, L. (2008) What's in a name? Commonalities and differences in public understanding of "climate change" and "global warming". *Public Understanding of Science* doi:10.1177/0963662506073088.

[7] The term 'climate change' means different things within the UN Intergovernmental Panel on Climate Change (IPCC) – climate change is irrespective of cause, natural or human – and within the UN Framework Convention on Climate Change (UNFCCC) – climate change is due solely to human causes.

[8] In Chapter 10: *Beyond Climate Change* I introduce lower-case 'climate change' and upper-case 'Climate Change' (see Box 10.1) to draw a further distinction between the physical and ideological connotations of the phrase.

not exclusive – cause of this change is human in origin. Where a more precise association of physical cause and effect is required, I preface climate change with the adjectives 'natural' or 'human-induced'/'anthropogenic'. This position is a compromise between the technical usage of the term by the Intergovernmental Panel on Climate Change – 'climate change irrespective of cause' – and the implicit popular association of climate change with causation due to human-induced emissions of greenhouse gases. This latter connotation is frequently shorthanded as 'global warming'; a practice also followed in this book. The technical term 'greenhouse effect' (now little in evidence in public discourse) retains its more limited meaning of describing the physical mechanism of the differential radiative heating of the atmosphere originally proposed by Joseph Fourier in 1824 and demonstrated experimentally by John Tyndall in 1859.

ONE

The Social Meanings of Climate

We are used to talking about summer heat in our poetry, but it is only when a real spell of it comes to us that we discover how rare it is. This July the whole countryside looks at the same time both strange and familiar. There is the corn, ripe as if it were the middle of August, and the dark foliage of later summer, but all our Northern landscape, unchanged in its forms and objects, is transfigured by the colours of the South. Usually, even in fine summer weather, there is a Northern coolness in our mornings and evenings; but now one is startled even in the early morning by the Southern splendour both of earth and sky (*The Times*, 26 July 1911).

The performance of the British climate over the past few months can at best be described as perfidious. After several very mild winters and two beautiful summers, including the most severe drought since records began 250 years ago, the climate has lurched to the other extreme ... the period from September 1976 until last June was the wettest for a hundred years (*The Times*, 19 August 1977).

So what can Britain expect as the blanket of greenhouse gases around the planet thickens? As the temperature nudged record levels last summer, the Met Office said that we should get used to such prolonged periods of settled, dry weather. There is a significant human contribution to these heatwaves because of carbon dioxide emissions over recent decades ... This is a sign of things to come ... Three years ago ... scientists ... showed that human emissions of greenhouse gases had more than doubled the risk of record-breaking heatwaves such as the one that is reckoned to have killed 27,000 people across Europe in 2003. By the 2040s, one summer in two is predicted to be hotter than 2003 (*Guardian*, 21 May 2007).

1.1 What is Climate?

We love our climate – and yet we fear it. As the above three interpretations of the climate of Britain reveal, we are not quite sure what to make of the idea of climate: we can celebrate its power to evoke strong emotions in us, while also bemoaning its unpredictability or fearing its future behaviour. We expect climate to perform for us; to offer us the weather around which we work and create and within which we relax and recreate. Yet we know too that climate is fickle, with a will and a mind of its own, offering us not only days of tranquillity and repose, but also the storms and dangers that our ancestors encountered over countless centuries and that continue to afflict us today.

Climate offers material benefits for all human cultures: the rain, wind, sun and warmth that waters, powers and feeds our lands and machines. Climate also offers resources for our aesthetic and spiritual imaginations: the clouds and sunsets which inspire our poetry, the seasonality around which we develop rituals. These benefits are often precarious, however, and this insecurity is a powerful driver of human innovation. New technologies, practices and systems are created to build social resilience in the face of a capricious climate. Constancy of climate is rare. Conversely, the precariousness of climate has also been invoked in explanations of the collapse of civilisations. Climatic stability has often been presumed to be a prerequisite for the stability of civilisations although, as we shall see later, the idea of climate change triggering societal collapse is itself not stable.

There may be 'good' or 'benign' climates and 'bad' or 'dangerous' climates, but only in the sense that climates acquire such moral categories through human judgements – judgements that suit our convenience or our capabilities. We do not judge climates against any fixed or universal morality. Is a 'good' climate a stable or a varying one? Is a 'bad' climate an unpredictable climate or one that is either too hot or too cold for our predilections? If you were going to design the ideal climate, what would it look like? All climates are difficult and yield dangers, yet all climates are fruitful and inspire creativity. There are few climates on Earth where humans have not lived and survived. Humans can accommodate a much greater range of the available climatic space than the ancient Greeks and early Medievals supposed. Sophisticated human civilisations are sustained in climates as dramatically different as that of 'torrid' Saudi Arabia (mean annual temperature 24°C) and 'frigid' Iceland (2°C) (see Figure 1.1). Yet there are few climates which, equally, do not carry danger or risk.

Since we are going to spend the next 300 or so pages exploring the reasons why we disagree about climate change, it is important that we dwell for a while on this idea of climate. Climate cannot be experienced directly through our senses. Unlike the wind which we feel on our face or a raindrop that wets our hair, climate is a constructed idea

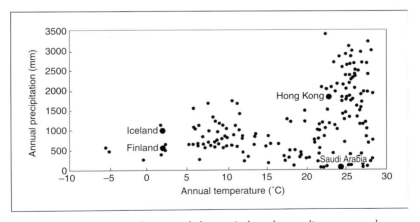

FIGURE 1.1: *The range of 'national climates' plotted according to annual average temperature (°C) and precipitation (mm) using statistics for the period 1961–90. Each dot represents the climate of one country.*

that takes these sensory encounters and builds them into something more abstract. Neither can climate be measured directly by our instruments. We can measure the temperature of a specific place at a given time, but no-one can directly measure the climate of Paris or the temperature of the planet. Climate is an idea that carries a much richer tradition of meaning than is captured by the unimaginative convention that defines climate as being 'the average course or condition of the weather at a place usually over a period of years as exhibited by temperature, wind velocity and precipitation'.[1] This chapter offers a guide to what this idea of climate means, using the insights offered by history, geography and anthropology.

Climate has both physical and cultural connotations. It has physical significance: one cannot deny that the climate of the Amazon is wetter in an absolute sense than is the climate of the Sahara. But climate also carries cultural interpretations: the climate of the Sahara means something quite different to a Bedouin than it does to a Berliner. We will explore these physical and cultural approaches to thinking about climate in Sections 1.2 and 1.3, respectively. Our ideas of climate may also carry more deliberate and entrenched meanings, being used to secure political or ideological goals. Ideas about climate are always situated in a time and in a place. As history gets rewritten and geography gets reshaped, so also change our ideas of climate. Climates can change physically, but climates can also change ideologically. Some of the ways in which the idea of climate has been a vehicle for promoting different ideologies, different ways of seeing the world, are explored in Section 1.4.

The idea of climate change, the subject of our book, is also an idea that has served many different purposes, and continues to do so. One of the most enduring of these is the way in which we have written about the human story using the language of climate change: the story of human evolution and the rise and fall of civilisations. Section 1.5 therefore offers a brief historiography of the different ways

---

[1] Merriam-Webster Collegiate Dictionary (1998) 10th edn.

in which we have written about climate change and human civilisation, about how we have frequently constructed antagonistic relationships between the vicissitudes of climate and the fates of nations and empires. The themes of the chapter, which reappear in various forms later in the book, are summarised in Section 1.6.

1.2 The Physical Basis of Climate

The idea of climate was first given linguistic form by the Greeks. The Greek word κλίμα, or *klima*,[2] was used as early as the sixth century BC by Pythagoras's disciple Parmenides to differentiate between five zones on the surface of the supposed spherical world. These latitudinal zones related directly to the inclination of the sun's rays on the Earth's surface, ranging from the torrid zone at the Equator to the frigid zone of the far North. These earliest attempts at climatic classification revealed the precariousness of the human relationship with climate. While the Greeks inhabited the forgiving temperate zone of the eastern Mediterranean, the frigid *klimata* of the North and the torrid *klimata* of the South were realms which gifted a legacy only of danger, or even death. Later Greeks further extended the idea of physical climates being dependent on latitude, and Ptolemy's seven *klimata* from the second century AD persisted as the conventional framework for explaining different climates well into the early Renaissance period.[3]

Attaching climate to latitude, to the inclination of the sun on the Earth's surface, lent a certain rigidity and constancy to the idea of climate which European explorers of the fifteenth and sixteenth centuries began to question. And as they did so, they raised wider questions about the authority of classical Greek science, as noted by historian

---

[2] Literally, 'slope' or 'incline'.

[3] Sanderson, M. (1999) The classification of climates from Pythagoras to Köppen. *Bulletin of the American Meteorological Society* 80(4), 669–73. Sanderson gives examples of world maps from the sixteenth century in which Ptolemy's seven climatic zones still offered the standard classification of known climates.

Craig Martin: 'The common experience of travellers to the New World … [showed] … that the theory of uninhabitable climatic zones was untenable and therefore [that] Aristotelian science was incomplete and fallible.'[4] Not only did Europeans survive the torrid and deathly climates of the Equator, they began to realise from far-off longitudes that latitude alone was a poor predictor of the climates they experienced. Enabled by the instrument revolution of the seventeenth and eighteenth centuries – which yielded barometers, thermometers and rain gauges – new ways emerged of understanding the physical and geographical attributes of climate.

The predominant means of capturing the physicality of climate was to be through meteorological measurement;[5] initially through individuals recording observations of the weather in private diaries and later, towards the end of the eighteenth century, through systematic and centralising networks of measurement. The application of standardised and regularised methods of observation of the natural world – one of the hallmarks of Western Enlightenment rationality – to what had previously been largely a philosophical or sensual endeavour, opened up new ways of describing climate and thinking about what it meant. Order was imposed on seemingly chaotic weather; first, by quantifying it locally at individual places and, subsequently, by constructing statistically aggregated climates from geographically dispersed sites. Climate for the first time became 'domesticated', revealing that, for example, British climate was 'generally temperate overall, but punctuated by bracing diurnal variations'.[6]

---

[4] p. 3 in Martin, C. (2006) Experience of the New World and Aristotelian revisions of the Earth's climates during the Renaissance. *History of Meteorology* 3, 1–16.

[5] There were other ways of capturing and describing the physical dimensions of climate: through its impacts on vegetation, phenology, ice cover, soil moisture. The Chinese had been particularly adept at recording such climatic indicators, some of them as far back as 1100 BC.

[6] p. 22 in Golinski, J. (2003) Time, talk and the weather in eighteenth century Britain, in Strauss, S. and Orlove, B. J. (eds), *Weather, climate, culture*. Berg: Oxford, pp. 17–38.

This quantification and standardisation of climate opened up new possibilities of interpretation and practical utility. Comparative climatic analysis could be undertaken, relying on numerical data rather than hearsay; an attractive prospect for nineteenth-century colonists and traders – how different was the climate of Cape Town from that of Amsterdam? And longitudinal studies of climate through time now became possible, providing a formal alternative to the reach of human memory – how stable really was climate?

As the nineteenth century began, this new way of describing climate through quantification of its physical attributes was gaining ground.[7] Standardisation of meteorological measurements was extending into the Americas and the tropical world, vigorously promoted by scientific entrepreneurs such as Alexander von Humboldt and the American meteorologist Matthew Maury, and the first systematic and quantitative large-scale climatologies were produced. In 1848, the Prussian physicist Heinrich Dove published the first global maps of monthly mean temperature, followed in 1883 by Austrian meteorologist Julius Hann's monumental *Handbüch der Klimatologie*. Its three volumes covered general, regional and local climates, and although Hann captured these climates primarily through the growing number of instrumental measurements, his third volume on local climates continued to use literary and eye-witness descriptions. The direct sensory and imaginative impacts of physical climate on the human mind were still seen as legitimate registers.

The quantitative and naturalistic approach to conceptions of climate found its ultimate expression in two of the most famous climatological products of the twentieth century. The Köppen classification

---

[7] For example, Clarence Glacken, in his 1967 book, *Traces on a Rhodian shore: nature and culture in Western thought from ancient times to the end of the eighteenth century*. University of California Press: Berkeley, CA, quotes (p. xv) Count Volney from his 1804 work on the climate and soil of the USA as remarking on the shift of meaning of the word 'climate', saying that 'the term climate is now synonymous with the habitual temperature of the air'.

of world climates, which Russian geographer Wladimir Köppen originated and refined between 1900 and 1936 and which is still in use today, marked the end of the transition from the original Greek classification of climate based on latitude. In Köppen's classification, the geographical complexities of regional climates are mapped by grouping together those climates whose statistical properties yield similar natural vegetation types. This leads to an infinitely more subtle arrangement of physical climates than imagined by Ptolemy.

A second icon of this physical approach to climate was first constructed only in the latter decades of the last century. The millions of individual thermometer readings taken around the world since the middle of the nineteenth century were compiled and synthesised into an index of an abstracted global climate – the globally averaged surface air temperature (Figure 1.2). This index of world climate – reconstructed back to 1850 and now routinely updated each month – both hides and reveals. It hides all of the heterogeneity of weather experienced in local

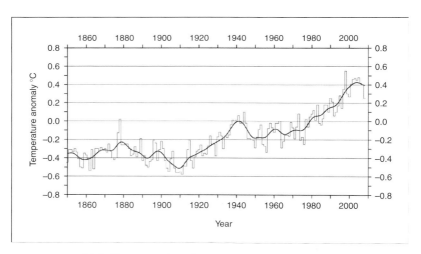

FIGURE 1.2: *Globally averaged surface air temperature for the period 1850–2008, expressed as anomalies from the 1961–90 average (°C). 2008 data are provisional.*
Source: Redrawn from Climatic Research Unit, UEA, website (accessed 9 July 2008).

places by local people and yet, by collapsing this diversity into a single numerical index, it reveals the behaviour of a large and complex global system. As we shall see later, this index has fulfilled many functions in the scientific and political discourses surrounding climate change; most importantly, perhaps, in lending simple and numerical visibility to the idea of climate (here measured as temperature) as an emergent property of an interconnected and physical global system.

The climatologists and meteorologists of the nineteenth century made the bravest attempts to reify climate, using a series of formal statistical rules to turn climate for the first time into an entity with quantitative description. This, of course, is how climate continues to be used in the physical and mathematical sciences, and opens up all sorts of possibilities for predicting future climate (in this physical sense). It is not surprising that, with its analytical roots so firmly planted in meteorology, the dominant popular understanding of climate therefore remains this numerical and statistical one. Thus the World Meteorological Organization insists that the climate of a place or region can only be robustly defined once it has been compiled from at least thirty years of meteorological measurement. Or to put it more pithily, 'Climate is what you expect, weather is what you get.'

The distinction between climate and weather remains one of the more elusive in popular discourse. While a degree of verbal ambiguity is appropriate for social intercourse, in analytical applications a more formal distinction becomes necessary. One way of visualising this distinction is shown in Figure 1.3, which uses the filter of time to demonstrate how we move between using descriptions of climate and weather depending on the relationship of the respective era to the present. The farther back in time we look, and certainly earlier than the last three or four centuries, the more our reconstructions of the past rely upon notions of climate rather than weather. Similarly, beyond the medium-range weather forecast, our descriptions of the future almost always reveal climatic categories rather than revealing weather events. On these distant past and future time-scales, the weather – the

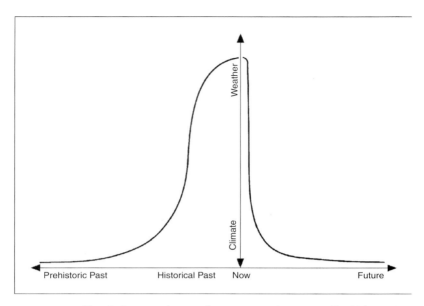

FIGURE 1.3: *Sketch diagram showing how we move between talk of 'climate' and talk of 'weather' depending on the relationship with the present. We can only access 'weather' for the next few days or for the past few centuries.*

minute-to-minute, day-to-day experience of the outcome of meteoro-logical processes – is largely hidden from us. We inevitably adopt the convenient shorthand of allowing climate to stand in for weather.

We have discussed the idea of climate thus far in predominantly physical terms. But the etymological origins of the word 'climate', and its subsequent attachment to aggregated meteorological measurements and eventually to the predictive natural sciences, only incompletely captures the subtlety and multiplicity of meanings with which the word 'climate' has been endowed. There have been many other ways of working with the idea we call climate; ways both less formal and more symbolic than those favoured by meteorologists. Thus climate may also mean 'the prevailing attitudes, standards or environmental conditions of a group, period or place',[8] a qualitative and less tangible

---

[8] On-line dictionary http://dictionary.reference.com/ [accessed 9 July 2008].

meaning than the statistical definition of physical climate employed by the World Meteorological Organization. Climate can be used to describe the prevailing milieu in politics, in the economy, or in a relationship – hence we talk about a region experiencing a 'climate of political unrest', about changes in the 'economic climate' of a nation, or find it used as literary symbolism, as in Nancy Mitford's 1949 novel *Love in a Cold Climate*.

These twin meanings – the meteorological and the metaphorical – of the English word 'climate' are also found in other languages; for example, Italian, Spanish, Thai, Chinese and Russian. In other languages, however, the idea conveyed by the world 'climate' is more complicated still. Anthropologist Timothy Leduc has explained the difficulties in translating the most relevant Inuktitut term '*sila*' into English. The Inuit of northern Canada use '*sila*' to talk about the 'weather' and 'climate' – the physical meteorology of the air – but also use the same word to describe the 'spirit of the air, a mystic power which permeates all of existence'.[9] In societies such as the Inuit, the separation of the idea of climate into distinct physical and cultural dimensions is linguistically prohibited. As explained by geographer David Livingstone, we therefore see that 'whilst ordinarily thought of as simply a constituent element of the natural order, climate has consistently surfaced as a cultural category'.[10] It is to this cultural category of meaning that we now turn.

1.3 The Cultural Basis of Climate

The ease with which we use 'climate' in its metaphorical sense helps us to understand that there are other ways of conceiving

---

[9] p. 237 in Leduc, T. B. (2007) Sila dialogues on climate change: Inuit wisdom for a cross-cultural interdisciplinarity. *Climatic Change* 85, 237–50.

[10] p. 79 in Livingstone, D. N. (2004) Climate, in Thrift, N., Harrison, S. and Pile, S. (eds), *Patterned ground: entanglements of nature and culture*. Reaktion Books: London, pp. 77–9.

climate's relationship with the physical world than solely through the meteorological statistics captured and reported by the World Meteorological Organization. Climate has always carried a deeper, precarious and more ambiguous meaning for humans than the merely prosaic.

Our biological evolution was forged through amplitudes of climate change – through dangerous encounters with climate – unknown to modern humans, while our cultural evolution has involved a variety of ways of mythologising and taming the workings of Nature's climate. The trail of the flood myth, for example, can be traced through many early cultures, most notably in the story of the Biblical Flood in the monotheistic religions of Judaism, Christianity and Islam. American cultural historian Clarence Glacken, in his 1967 masterpiece *Traces on a Rhodian Shore,* observed how the idea of climate has been used in Western thought to explain elements of human character, activity and culture. Hippocrates, in the fourth century BC, was the earliest to develop such an association through interweaving the histories of medicine, geography and anthropology in his treatise on *Airs, Waters, Places.* In this Hippocratic sense, the physicality of climate acts as a constraint on human action and as a shaper of human fortunes. The intimacy of relationship between culture and climate is nowhere better illustrated than in the case of Egypt and the Nile. The climatic pulsing of the river through annual and seven-yearly cycles gave – and still gives – life, sustenance, shape and meaning to Nilotic cultures.

The idea of climate also interacts with the human psyche and with cultural practice in less material and more imaginative ways. Climate is frequently bound up in our sense of national identity, threading its way into many aspects of our social memory, while climatic extremes are adopted as anchors for personal memory in both modern and traditional societies. There are few inhabitants of New Orleans from August 2005, for example, who will not end their lives being able to relate events in their life story to either pre-Katrina or post-Katrina eras. The human experience of climate releases powerful emotions which can be both benign and threatening. Various studies have shown correlations between climatic fluctuations and, for example, sexual activity or suicide rates. Indeed, throughout the human experience of realised climate and our anticipation of portended climates, there runs a thread of anxiety and fear: 'The history of humanity is characterised by an endemic anxiety … it is as if something or someone is remorselessly trying to sabotage the world's driving force – and particularly its climate.' [11] The persistent use of visual icons of glaciers and palm trees – ice sheets or deserts – as signifiers of climatic danger reveals such anxiety.[12]

As we have observed earlier, until the eighteenth century climate was largely unquantified – and unquantifiable. Weather and climate were described in qualitative and impressionistic terms, expressing beauty or prosperity, danger or threat; the latter commonly carrying portentous meaning. Experiences of extreme weather – as of other natural phenomena such as earthquakes – have long been interpreted by individuals and cultures as signifiers of divine blessing or judgement. The relationship between God and climate, especially drought, portrayed in the early Jewish scriptures, for example, reveals this particular reading of weather extremes. A theological interpretation of the capriciousness of climate remained dominant in western Europe through the later Middle Ages and well into the early modern period. It remains a common frame today in many traditional cultures.[13]

The fears evoked by unprecedented extremes of weather were fuelled by a particular understanding of the relationship between God

---

[11] p. 149 in Boia, L. (2005) *The weather in the imagination* (trans. R. Leverdier). Reaktion Books: London.

[12] For evidence of such iconography in European literature and media reporting, see Brönnimann, S. (2002) Picturing climate change. *Climate Research* 22, 87–95.

[13] And also in some supposedly non-traditional societies. In July 2007, the Anglican Bishop of Carlisle claimed that the extensive flooding in parts of northern and central England caused by heavy rains were a judgement from God: 'We are reaping the consequences of our moral degeneration, as well as the environmental damage that we have caused.' (*Sunday Telegraph*, 2 July 2007).

and Nature. Weather was beyond human understanding or control and was seen as a primary instrument for the exercise of God's expressions of favour or disfavour on morally vulnerable populations. These fears could be augmented by a parallel demonising of the causes that lay behind adverse climatic experiences. Seemingly without rational cause, climate and weather were viewed as the territory within which both divine and satanic influences were at work. Through the attachment of extremes of weather to divine retribution or demonology, climate frequently remained associated with fear through the late Middle Ages and early modern periods.

Although the theological language may have changed, Western societies continue to be attracted to the idea of moralising the performance of climate, or at least to moralising the driving forces that lie behind climate. British scientist Jim Lovelock's original idea of personifying the biophysical Earth system using the metaphor of Gaia – the Earth Goddess – has had a profound influence on some of the ways in which science and society have come to view climate and to speak about it. Rather than being a function of God's displeasure, extreme climatic behaviour can now be interpreted as Gaia's revenge on a morally wayward and abrasive humanity. Thus the eminent American oceanographer Wally Broecker is comfortable using the metaphor of an 'angry beast' to describe climate change – 'the climate system is an angry beast and we are poking at it with sticks'[14] – reminiscent of many non-Western forms of personifying Nature. Secular culture is reinterpreting a more ancient human instinct.

As well as describing a physical reality, climate then can also be understood as an imaginative idea – an idea constructed and endowed with meaning and value through cultural practice. Registers of climate can be read in memory, behaviour, text and identity as much

---

[14] p. 7 in Broecker, W. S. (1999) What if the conveyor were to shut down? Reflections on a possible outcome of the great global experiment. *Geological Society of America Today* 9(1), 1–7.

as they can be measured through meteorology, as science historian Jan Golinski shows in his enlightened survey of eighteenth-century British attitudes to weather and climate: 'This ... view of the climate was bound up with the sense of British national identity.' [15] A cultural reading of climate is also necessary to understand the original meaning of climatic phenomena such as El Niño in the Pacific Ocean. El Niño – literally 'the boy' or 'the small one' – was the name given by Roman Catholic Peruvian fishermen in the nineteenth century to a warm coastal current that appeared sporadically along the shores of Ecuador and Peru. By bringing welcome relief from the cold waters that usually prevailed and by watering the arid coastal deserts, and because this warm current usually appeared around Christmas-time, the climatic phenomenon became identified with the character of Jesus Christ, a natural blessing associated with the Saviour of the World.

I want to suggest that the separation of the cultural from the physical that is implicit in the now dominant understanding of the idea of climate is a peculiarly Western separation. It is one which gained ascendancy as the Enlightenment project unfolded in Europe and as its ripples were felt further afield. The separation of Culture from Nature, illustrated above in the way we have presented the idea of climate as either physical or cultural, has been suggested by the French sociologist Bruno Latour to be one of the essences of modernism. Such dualism was rarely found in traditional non-Western cultures and yet exposure of such cultures to Enlightenment rationality has influenced the way in which climate is perceived. As anthropologist Julie Cruikshank observes: 'The [Western] idea that a measurable natural world might be prised from its cultural moorings has continued to insinuate itself in locations and landscapes where local understandings were conventionally framed very differently.' [16] In such locations and landscapes,

[15] p. 57 in Golinski, J. (2007) *British weather and the climate of enlightenment.* University of Chicago Press.
[16] p. 245 in Cruikshank, J. (2005) *Do glaciers listen? Local knowledge, colonial encounters and social imagination.* UBC Press: Vancouver, Canada.

the traditional understanding of climate was instinctively and deeply loaded with cultural meaning, as illustrated in Box 1.1 describing early Icelandic colonists. Evidence of such holistic approaches to thinking and living with climate can still be found today; for example among African pastoralists and the Canadian Inuit.

Box 1.1: Climate in Icelandic Sagas[17]

The Sagas of the Icelanders were written in the thirteenth and fourteenth centuries and recount stories of prominent Icelandic families during the early settlement period of Iceland by the Vikings around the tenth century AD. These Sagas, a unique genre of medieval literature, reflect the way in which Icelanders thought and talked about the climate and their relationship to it. The natural land- and seascapes were so closely entwined with their cultural expressions that one cannot separate Nature from Culture. In medieval Icelandic society, as with most traditional or pre-modern societies, climate was an integral part of Culture. One could not observe and analyse climate dispassionately from outside; one could only experience and interpret climate from inside the culture.

The sea and sailing were central to the lives of these Vikings. Sea traffic between Iceland, Norway and Greenland and the New World was vital for subsistence and communication. The Vikings had an intimate knowledge of the ocean climate – sea-ice, storms, winds – based on careful observation handed down from generation to generation. The climates encountered on the open ocean were respected as sentient beings, as the gods Aegir and Hler, who

---

[17] Box 1.1 draws upon material in Ogilvie, A. E. J. and Pálsson, G. (2003) Weather and climate in Icelandic sagas, in Strauss, S. and Orlove, B. S. (eds), *Weather, climate, culture.* Berg: Oxford, pp. 251–74.

could act as both givers and takers of life. The medieval Icelandic economy was based on livestock farming and the annual hay harvest was central to survival. The Sagas contain many references to the abundance or otherwise of summer harvests and the harshness of winter conditions. For subsistence farming and fishing in harsh environments, the reliability of climatic knowledge and weather signs was often critical. Acute observation and respect for the cultural and economic signifiers of changes in climate were central to these Icelandic conceptions of climate itself.

Climate was also used metaphorically in the Icelandic Sagas to suggest powerful human emotions. Thus the dark thoughts of the mind were likened to 'dark mists that are drawn up out of the ocean', while the brightness of a woman's warm and gentle love compared to 'sunshine and gentle weather'. Changes in weather were also used to embody meanings hidden in the narrative. In the story of two warrior friends, Thorgeir and Thormod, the reader becomes aware that 'the day has been bright with sunshine but the weather is now starting to thicken' signifies that events in the human world are also about to turn unpleasant.[18]

The difference in the approach to climate between the Western rationalist tradition and that implicit in much local environmental knowledge is a difference we will dissect in more detail in later chapters. For now we need to explore a further aspect of how the idea of climate is entangled in the human cultures which experience and talk about it. We need to understand the ways in which the idea of climate is often called upon to act as a justification and conveyor of ideology.

[18] *Ibid.*, p. 265.

1.4 Climate as Ideology

The classical Greek idea of climate served several purposes. Climates to the far north were too cold and those in the far south too hot to allow significant human cultural development. The climates of the intermediate Mediterranean zone, on the other hand, were just perfect for the blossoming of human creativity, expressed through culture, wealth creation and innovation. This circularity of reasoning was therefore a way of simultaneously explaining and justifying Greek hegemony in the classical world.

Jump forwards to our own century and we find that contemporary natural science conceives climate very differently. Rather than differentiated climates in particular places, the dominant intellectual frame is to see climate as a globally functioning holistic system; the totality of the atmosphere, hydrosphere, biosphere and geosphere, together with their interactions. Climate now connotes 'less the weather of any particular place than something more closely akin to the global environment: a natural object to be understood, investigated and managed on planetary scales'.[19] And this investigation and management requires the skills of the planetary scientist. This conception of climate is different from the Greek *klimata*, but the end result is the same. Climate as an idea simultaneously explains and justifies – in this case – the hegemony of the Earth (or climate) system scientist.

The idea of climate has been changing as much as, if not more than, the physical climate itself. Climate has been as much a carrier for ideologies[20] in the past as it continues to be in the present. We will discover evidence of the different ideological burdens borne by climate as we move through the book, but for now let us briefly note four

---

[19] p. 7 in Miller, C. A. and Edwards, P. N. (eds) (2001) *Changing the atmosphere: expert knowledge and environmental governance*. MIT Press: Cambridge, MA.

[20] By 'ideology' is meant 'the body of doctrine, myth, belief, etc., that guides an individual, social movement, institution, class, or large group'. Marxism and racism are obvious examples of ideologies; science and religion more controversially so.

of the more interesting ones that have revealed themselves: racism, mastery of Nature, the wildness of Nature, and system (in)stability.

Racism

The idea that the character of different races is shaped, or even determined, by climate has been one of the more enduring in the intellectual history of climate. We have already alluded to the way in which Greek conceptions of climate 'fortuitously' lent the Aegean Sea the most hospitable climate of the classical era. Herodotus, in the fifth century BC, could therefore understand the difference between Egyptians and Greeks: 'In keeping with the idiosyncratic climate which prevails there and the fact that their river behaves differently from any other river, almost all Egyptian customs and practises are the opposite of those of everywhere else.'[21] In similar vein, Hippocrates could 'explain' the triumphs of Alexander over the Persian Empire in terms of contrasting climate-moulded racial character; the energy of Europe winning over the 'softer' character of Asia.[22]

Similar reasoning can be found in Arab literature from the fourteenth century – the Tunisian geographer Ibn Khaldun – and in the writings of late Renaissance and early modern European thinkers – Jean Bodin, Baron Montesquieu and David Hume. The father of the Enlightenment, German philosopher Immanuel Kant, could therefore remark in 1775: 'The inhabitant of the temperate parts of the world, above all the central part, has a more beautiful body, works harder, is more jocular, more controlled in his passions, more intelligent than any other race of people in the world. That is why at all points in time these peoples have educated the others and controlled them with weapons.'[23]

---

[21] Herodotus, *The histories* (trans. R. Waterfield, Oxford, 1998), Book 2, p. 35, cited in Boia (2005), *The weather in the imagination*, p. 23.

[22] Boia, *The weather in the imagination*, p. 25.

[23] From Kant, I. (1775) *On the different races of Man*, cited in Livingstone (2004) Climate, p. 78.

The most developed articulation of this line of reasoning, of using a conception of climate to support a racist ideology, came from the American geographer Ellsworth Huntington in the early decades of the twentieth century. Drawing upon a number of empirically based studies examining the productivity of factory workers under different climatic conditions Huntington reaffirmed what Hippocrates, Ibn Khaldun and Montesquieu would also have 'known', namely that 'the denizens of the torrid zone are slow and backward, and we almost universally agree that this is connected with the damp, steady heat'.[24] Huntington's views, and those of other early twentieth-century geographers and philosophers, strike us today as naïve and even dangerous, and yet the deterministic philosophy underpinning such thinking can still lurk near the surface. Thus a 1993 analysis of the relationship between climatic characteristics and political stability concluded that 'it may never be possible to prove absolutely that a mild climate in mid-latitudes helps to foster a tolerant society or that an extreme climate may predispose people towards intolerance … however, the historical record is highly suggestive [of such a link]'.[25]

William Meyer, in his fascinating survey of *Americans and Their Weather*, coins the term 'meteorological fundamentalism'[26] to rebuke those who implicitly or explicitly assume that the significance of climate can be established purely from its physical characteristics without regard for cultural conditioning or human agency. Traces of such climatic fundamentalism can be found in some of the current rhetoric around the idea of wars, conflicts and refugees induced by climate change. Many have claimed, for example, that the violence and civil

---

[24] p. 2 in Huntington, E. (1915/2001) *Civilisation and climate*. University Press of the Pacific: Honolulu, HI.

[25] pp. 63–4 in Beck, R. A. (1993) Viewpoint: climate, liberalism and intolerance. *Weather* 48, 63–4.

[26] p. 71 in Meyer, W. B. (2000) *Americans and their weather*. Oxford University Press.

war in Darfur, western Sudan, during recent years has been largely driven by changes in climate: 'Darfur remains a dramatic example of how a small shift in climate can have dramatic and horrifying human consequences.' [27]

Mastery of Nature

Ellsworth Huntington's legacy can also be traced in the way in which climate has been used to reveal and propagate the ideology of human mastery of Nature, specifically the mastery of climate. Huntington ended his 1915 book, *Civilisation and Climate*, with these words, 'If we can conquer climate, the whole world will become stronger and nobler.' [28]

The idea that climate is a physical manifestation of the natural world that is available for 'conquering' is recurrent throughout the intellectual history of the West. Clarence Glacken's survey of 2,000 years of Western environmental thought[29] adopts the idea of human modification and control of Nature as one of the three enduring themes throughout this period of Western history, although it is only in the last 200–300 years that humans – either through technology or through sheer weight of numbers – have realised the full extent of their influential abilities. Climate has been the object of many attempts at 'conquering' – whether of local, regional or now global climates – and the deep-seated appeal of this ideological project is suggested by the diverse cultural and political settings in which these attempts have been launched. Box 1.2 gives examples from such unlikely ideological companions as Victorian imperialism, American capitalism and Soviet communism.

[27] p. 48 in Walker, G. and King, D. (2008) *The hot topic: how to tackle climate change and still keep the lights on.* Bloomsbury Press: London. This claim is refuted by: Kevane, M. and Gray, L. (2008) Darfur: rainfall and conflict. *Environmental Research Letters* 3 doi:10.1088/1748-9326/3/3/034006 10pp.

[28] Huntington, *Civilisation and climate*, p. 294.

[29] Glacken, *Traces on a Rhodian shore.*

Box 1.2: The Conquering of Climate[30]

Victorian Imperialism. Danger surrounded the Victorian conception of tropical climates. Whether due to degeneracy, depravity or debility, the encounters with the unknown climates of southern Asia, Africa and South America by white settlers evoked fears and anxieties about climate that emerged from the imperial ideology of the time. These newly encountered climates took on the roles ascribed to them by the prevailing and dominant culture. The moral classification of tropical climates as dangerous and threatening was tightly bound up in the discourse around acclimatisation – could white Europeans settle, survive and rule in 'hostile' climates? The association of (tropical) climate with fear, danger and anxiety was as much a function of the imperial ideology of the day as it was a function of detached physical or medical diagnosis. Opinion became polarised in the later Victorian period about whether or not the unknown and forbidding climates of the tropics were to be feared, and thus were in need of 'conquering'.

Tropical climates *were* conquered, in a literal as well as a metaphorical sense. As the European imperial adventure lost its way in the twentieth century, the psychological hold on the European mind of the pathology of tropical climates was dissipated. It was widely regarded that sustainable colonisation of India by Europeans required periodic escape by the settlers to the cooler climate of the Indian hills, driving the construction of hill stations as white enclaves. And later improvements in tropical medicine and air-conditioning technologies removed some of the direct physical fears

---

[30] The sources for Box 1.2 are: Fleming, J. R. (2006) The pathological history of weather and climate modification: three cycles of promise and hype. *Historical Studies in the Physical and Biological Sciences* 37(1), 3–25; and Hulme, M. (2008) The conquering of climate: discourses of fear and their dissolution. *Geographical Journal* 174(1), 5–16.

that tropical climates presented to non-indigenous populations; an outcome foreshadowed by Huntington in 1915 using the idiom of the era. 'In the future we can scarcely doubt that this method of overcoming the evil effects of a tropical climate will be resorted to on a vast scale, not only by foreigners, but by the more intelligent portion of the natives.'[31]

American Capitalism. Another route to the conquering of climate lay through scientific weather modification, a practice first advocated in the United States of America during the nineteenth century. Burning large tracts of forest to accelerate rain-making thermal currents was one of the first methods tried in 1849 by the meteorologist James Espy (also known as 'The Storm King'), although success eluded him. A later attempt by a retired general from the American Civil War, General Daniel Ruggles, and a Washington patent lawyer, Robert St. George Dryenforth, used the so-called 'concussion theory' of rain-making to induce rain in the Texan outback. The US Congress allocated $10,000 for this experiment in 1891 in which large explosions in the atmosphere were triggered in an attempt to keep the weather in an unsettled condition. No lucrative patent was forthcoming.

These idiosyncratic attempts at conquering climate through bringing the weather under human control presaged more serious and scientifically well-founded efforts later in the twentieth century to modify rainfall through cloud-seeding. The scientific entrepreneur Irving Langmuir was the great enthusiast for this technology, claiming that seeding clouds with dry ice or silver iodide opened possibilities for redirecting hurricanes, generating artificial snow storms, and watering the arid lands of Arizona. Although this technology has not promised all that Langmuir had hoped, it continues to attract investment today. In 2007, the

---

[31] Huntington, *Civilisation and climate*, p. 291.

Chinese reported a successful attempt to create snowfall over the city of Nagqu in Tibet by seeding clouds with silver iodide. 'This proves it is possible to change the weather on the world's highest plateau.'[32]

Soviet Communism. The desire to conquer climate reached its apogee during the 1960s in Soviet Russia. In keeping with Stalin's Great Plan for the Transformation of Nature, Russian authors, visionaries and engineers imagined numerous ways to bring Russian climate, if not world climate, under centralised control. Schemes to dam the Bering Straits to melt Arctic ice and warm northern Russian shores; to flood the Qattara Depression in Egypt to bring rain to the Sahara; or to divert the Gulf Stream with a dam between Florida and Cuba to enhance the northward flow of warm water all featured in the large imaginations of these Soviet entrepreneurs. Climate was to become another site of ideological conflict between West and East as the communists saw the diversion of the River Congo into the northern dry heartland of Africa as a way to counter growing American economic imperialism in Africa.

And it was in part reaction to this ideological battle, combined with the first inkling of worldwide changes in climate induced by rising concentrations of carbon dioxide in the atmosphere, that the USA first endorsed their own rationale for large-scale climate control. In 1965, President Johnson's Science Advisory Committee suggested that 'the possibilities of deliberately bringing about countervailing climatic changes ... need to be thoroughly explored'.[33] More than forty years later their wish is now being fulfilled.

---

[32] p. 957 in *Nature* 446, 26 April 2007.

[33] p. 127 in President's Science Advisory Committee (1965) *Restoring the quality of our environment.* Report of the Environmental Pollution Panel (November), Washington DC.

This ideological project to conquer and control climate continues today at both a micro and a macro level. Most new cars now sold advertise their ability to 'climate control' the ambient air temperature inside the vehicle. By engineering the exchange of air and heat into and out of the body of the car, (micro-)climate can be brought under human mastery. 'Climate control' is also a project on the planetary scale. An increasing number of schemes have been put forward to geo-engineer the global climate, to manipulate the physical function-ing of the Earth's system in such a way as to achieve a stated climatic goal or to limit a feared climatic disturbance. Injecting sulphur into the stratosphere, sucking up biologically productive cold water from the deep oceans, building carbon dioxide extraction machines in the free atmosphere, are all suggestions revealing latter-day attempts to master and conquer climate. That such proposals are advocated in response to *our* inadvertent modification of world climate – and thus to 'limit global warming' to some desirable level – does not weaken the argument. The point is that the enduring idea of climate mastery reveals an ideological position that humans have the desire, the right and the means to control the climatic forces of Nature.

The wildness of Nature

This brings us to a third example of how climate has been used to carry diverse meanings – the ideology of the wild. In direct contrast to the claims and goals of mastering and controlling climate, this dis-course sees climate as a repository of what is natural, something that is pure and pristine and (should be) beyond the reach of humans. Climate therefore becomes something that is fragile and needs to be protected or 'saved', just as much as do natural landscapes or animal species. These are goals which have fuelled the Romantic, wilderness and environmental movements of the Western Enlightenment over two centuries or more. Of course it can be argued about when and why Nature first became a separate category in the human imagination, or indeed whether the category of wildness exists in any substantive sense. There is no denying, however, that the idea of wildness has been a persistent mode of discourse over recent centuries and that by losing wildness, by taming or mastering Nature, humans are diminishing themselves and maybe something beyond themselves.

Bill McKibbin, in his classic book *The End of Nature*, gives powerful voice to this sentiment with regard to climate. His lament for the end of Nature finds its highest expression in the transition from a natural climate to a climate which is being modified through human interference with the global atmosphere. That 'a child will now never know a natural summer'[34] is for McKibbin a cause of sadness and of loss. That global climate is no longer safe from the contaminating influences of the human species speaks symbolically of just how deeply humans have penetrated the idea of the natural.

System (in-)stability

And it is perhaps this ideology of wildness, the idea of world climate as the ultimate refuge of the natural, which has driven much of the thinking which lies behind the fourth example I use of how we load climate with our ideologies. If an untouched climate, a pure and natural climate, is to be valued, then maintaining its stability becomes of prime, even sacrosanct, importance. Climate thus becomes freighted with the ideology of stability and order in Nature, as opposed to ideas of change and chaos.

The ideology of stability fits very well the modern scientific conception of climate as a outward physical manifestation of the workings of a global system: local climates are delivered through the functioning of an intricate planetary system, a system which at a metaphorical level might even have sentience and teleology. James Lovelock's metaphor of the Earth as a self-regulating organism – as Gaia, Mother Earth – lends itself to valuing stability and order within the climate system. It is perhaps for this reason that one discourse around climate

---

[34] p. 55 in McKibbin B. (1989) *The end of nature*. Random House: London.

change views a stable climate as a public good; a view counterposed against the outcome if humans are allowed to mess with Gaia, namely climate chaos: 'This fever of global heating brought on by a plague of people is real and deadly and might already have moved outside our and the Earth's control.'[35]

   This pejorative phrase – climate chaos – has been adopted by many climate change campaigners in recent years, although it is an ironic adoption since the physical sciences have revealed that the functioning of the atmosphere is in fact naturally chaotic.[36] In this reading, climates – from local to global – become entities that must be stabilised for the public good, an idea enshrined in Article 2 of the UN Framework Convention on Climate Change signed in 1992: 'The ultimate objective of this Convention … is to achieve *stabilisation* of greenhouse gas concentrations in the atmosphere' [emphasis added]. Change, dynamism and instability – foundational attributes we see evident in the natural functioning of most ecosystems – are to be resisted when discovered or created in the climate system. 'Tipping points' are the new elemental dangers that must be avoided, and (re-)stabilising global climate following the reckless interference of humans is what must be achieved at all costs.

These four examples – and there are many more that could be articulated – of how the idea of climate carries and promotes a range of ideological projections have been introduced here for just one reason; a reason that is central to this book. These examples should be enough to convince us that describing climate is about much

---

[35] p. 1 in Lovelock, J. (2006) *The revenge of Gaia*. Penguin: London.
[36] The late American mathematician and meteorologist Edward Lorenz (1917–2008) pioneered understanding in the early 1960s of the atmosphere as a chaotic system, publishing the first exposition of his ideas in 1963. It has since become an essential theoretical insight for all meteorological, and increasingly climatic, prediction.

more than describing some physical attribute of the planet we live on. The idea of climate exists as much in the human mind and in the matrices of cultural practices as it exists as an independent and objective physical category. The multiple meanings of climate, and the ideological freightage we load onto interpretations of climate and our interactions with it, are an essential part of making sense of what is happening around us today in our climate change discourses. Understanding climate – and hence understanding climate change – as an overly physical phenomenon too readily allows it to be appropriated uncritically in support of an expanding range of ideologies beyond the four summarised here; for example the ideologies of green colonialism, of the commodification of Nature, of national security, of celebrity culture. I am not judging whether or not any of these creeds are desirable: some may be, some may not. The more important point is that by disconnecting climate from its cultural forms – by framing climate as overtly physical and global – we allow the idea of climate change to acquire a near infinite plasticity. Detached from its cultural anchors, climate change becomes a malleable envoy enlisted in support of too many rulers.

Before we leave this chapter on the social meanings of climate, we need to observe one final important facet of the climate story, one that emerges from recognising the multiple ideologies the idea of climate can carry. This final section observes the different ways in which *changes* in climate have been used to tell stories about the rise and fall of human civilisations.

1.5 Climate in History

Changes in climate have been invoked to (help) explain the decline of civilisations more often than they have been interrogated to explain their rise. Edward Gibbon's classic account of the decline and fall of the Roman Empire was the first systematic attempt to relate climatic factors to the declining fortunes of a major civilisation. Although only a part of his story, Gibbon depicted the confrontation between Romans and the threatening German hordes in climatic terms and also hinted at the climatic origins of the declining agricultural yields in Rome's Mediterranean bread basket, which weakened the agrarian pillar of the Empire. Gibbon's account, published in the revolutionary period between 1776 and 1788, was the first in the genre, but it is a genre that has multiplied in the following two centuries.

For Ellsworth Huntington it only took a short step from his climatic theory of racial hierarchy to arrive at a climatic basis for the shifting fortunes of the classical civilisations of the Near East, the Mediterranean and Central America. In the same 1915 book in which he 'demonstrated' the racial superiority of whites over blacks, he devoted two chapters to exploring the idea that the fate of civilisations is intimately related to changes in climate. He concluded that, 'No nation has risen to the highest grade of civilisation except in regions where the climatic stimulus is great … a favourable climate is an essential condition of high civilisation.'[37] The corollary of this reasoning is that if the climatic stimulus is withdrawn or weakened, decline will inevitably follow. Huntington's intellectual legacy endured through many decades of the twentieth century, even if not always expressed quite as stridently. The British climatologist Charles Brooks echoed Huntington's logic with respect to the decline of the Indus, Mesopotamian and Egyptian civilisations in his 1926 book *Climate Through the Ages*, while Austin Miller's standard textbook *Climatology*, which went through nine editions from 1931 onwards, was still claiming in 1961 that '[civilised] power has followed the cyclonic belt in its northward retreat'.[38]

More sophisticated and nuanced arguments about the role of climate in the collapse of civilisations emerged during the 1970s. Thus Reid Bryson and Thomas Murray, in their popular study *Climates of*

---

[37] Huntington, *Civilisation and climate*, p. 270.
[38] p. 305 in Miller, A. A. (1961) *Climatology* (9th edn). Methuen: London.

Hunger, eschewed the broad sweeping generalisations of Huntington and Brooks in favour of detailed case studies of how sustained episodes of drought had undermined the societies of Mycenae in Greece, the First Nation Indians of the American Great Plains, and the Malian Kingdom in West Africa. Similarly detailed archaeological and palaeoclimatic work has suggested that the collapse of the Mayan civilisation around the ninth century AD was triggered by a severe drought. Archaeologist Harvey Weiss claims that 'many lines of evidence now point to climate forcing as the primary agent in repeated societal collapse'.[39] The recent narrative stories told by Brian Fagan[40] and Jared Diamond[41] continue this tradition of seeing the hand of climate change as a significant destabilising factor for human civilisations. Diamond, for example, while not necessarily promoting a simple determinist account of human civilisations, presents climate change as one of his five factors that drive societal failure.

This discourse about the societal perils of climatic instability remains dominant. Civilisations evolve in benign environments, but through poor resource management – often exacerbated by climatic change – societal failure follows. The most extreme expression of such thinking can be found in Jim Lovelock's *Revenge of Gaia*. Lovelock sees the future survival of *all* human civilisation being dependent upon restabilising the global climate. In his view, civil structures and institutions are so intricately bound to the climate of the twentieth century that any sustained excursions from this optimal climatic state spell disaster: 'We [humans] are now so abusing the Earth that it may rise and move back to the hot state it was in 55 million years ago and if it does most of us and our descendants will die.'[42]

---

[39] p. 610 in Weiss, H. and Bradley, R. S. (2001) What drives societal collapse? *Science* 291, 610–11.

[40] Fagan, B. (2003) *The long summer: how climate changed civilisation*. Basic Books: New York.

[41] Diamond, J. (2005) *Collapse: how societies choose to fail or succeed*. Penguin: London.

[42] Lovelock, *Revenge of Gaia*, p. 1.

Yet the historiography of climate change and civilisation has another, opposite, tone which offers a different dynamic, or at least a deeper contingency, than these conventional readings. The logic of Huntington, Bryson and Diamond is inverted such that climate change acts as a stimulus *for* innovation and societal adaptation, a stimulus that – rather than threatening a civilisation – can accelerate the development of new complex civil and social structures. The British historian Arnold Toynbee favoured this view when he claimed in the 1930s that 'ease is inimical to civilisation … the greater the ease of the environment, the weaker the stimulus towards civilisation'.[43]

This contrary (and admittedly minority) reading sees civilising processes being a reaction to conditions of climatic stress and deterioration: benign or stable climates would not provoke the necessary innovation or adaptation through which new social forms come into being. Sustained regional drought and aridification, for example, rather than undermining social stability, could act as a trigger for increased social complexity associated with urbanisation and state formation.[44] This reading of the evidence also sits with more nuanced views that argue that the outcome of any given climatic shift is deeply dependent on the resilience and adaptability of local societies and institutions. There is no simple mapping between climate change and the fortunes of civilisations.

It is also worth observing that in a different context – that of human biological development – the fluctuations of climate over hundreds of thousands or millions of years have acted as evolutionary stimuli. There is substantial evidence in support of the idea that climate-driven changes in the environment over these evolutionary time-scales were responsible for hominid speciation, enlarged cranial capacity, and cultural

---

[43] Toynbee, A. J. (1934) *A study of history. Vol. II: The genesis of civilizations*. Oxford University Press.

[44] This is an argument explored in Brooks, N. (2006) Cultural responses to aridity in the Middle Holocene and increased social complexity. *Quaternary International* 151, 29–49. He cites evidence from the Near East, the Sahara, South Asia, coastal Peru and northern China in support of this position.

innovations.[45] The relationship between climate change and humanity, traced through its biological and social development, is a long one.

1.6 Summary

This chapter has introduced climate as an idea. It is an idea that can be approached using either physical or cultural pathways, but it is best understood as an idea that binds together the physical world and our cultural imagination. The idea of climate originated deep in the human past and is one around which our notions of Nature, Culture and history have formed and re-formed. Climate is therefore not just an abstract idea, but also a somewhat elusive one; a bit like 'goodness' – easier to recognise than to define.

Four themes have been introduced in this chapter, echoes of which will be found later in the book as we examine why we disagree about climate change. The first theme is that climates have both physical reality and cultural meaning. Indeed, the idea of climate can only be fully understood if one allows these physical and cultural dimensions to interact and mutually shape each other. Treating climate purely as a physical entity, accessible solely through natural science or, conversely, allowing the cultural symbolism of climate to be detached from any physical anchors, denies something essential about the idea of climate. Following from this is the second theme – climate is frequently used to carry and to convey a variety of ideological assumptions and projections. This ideological baggage may not always be obvious at first sight. The examples given in this chapter show how climate has been used to support, *inter alia*, the ideologies of racism, the human mastery of Nature, the sanctity of a pristine Nature, and the preference for stability over change. The idea of climate, transcending the physical and the cultural but drawing substance from both, lends

---

[45] These arguments are summarised in Behrensmeyer, A. K. (2006) Climate change and human evolution. *Science* 311, 476. To this list of events she adds the morphological shift to bipedality, behavioural adaptability and intercontinental migration.

itself to such multiple, layered, and sometimes contradictory readings. The third theme is that climates change over time – both the physical climates of places, but also the ideologies with which climate is associated. Climate may change because its physical attributes change or because its cultural symbolism changes, or both.

And finally, we need to recognise that the ways in which the story of climate change and human civilisation has been told have also changed over time. The dominant trope in this story has been one of climate change as threat, and yet dissenting voices have emerged which emphasise the creative potential for societies that can be found through changes in climate. We cannot detach the stories we tell about climate from the stories we tell about societies. And if this is true for climate, then it holds additional truth in relation to climate change: 'How [and why] climate change matters will always depend upon how society has evolved and continues to evolve, just as will the significance of those aspects of climate that remain the same.'[46] Disagreements about climate change are as likely to reveal conflicts within and between societies about the ideologies that we carry and promote, as they are to be rooted in contrary readings of the scientific evidence that humans are implicated in physical climate change.

We next need to examine what this evidence is and how it has emerged over the last 150 years. This will be our task in Chapter 2: *The Discovery of Climate Change*.

FURTHER READING FOR CHAPTER 1

Boia, L. (2005) **The weather in the imagination.** Reaktion Books: London.
Written by a Romanian historian, this delightful book traces the cultural history of the idea of weather and climate from the Greeks to the present day. The idea of climate and climate change has appeared in religion, in science, in society and in politics in different ways and at different times. Lucien Boia argues that the imaginative role of the mind is as important as the physical experience of the body.

[46] Meyer, *Americans and their weather*, p. 213.

Glacken, C. (1967) **Traces on a Rhodian Shore: nature and culture in Western thought from ancient times to the end of the eighteenth century.** University of California Press: Berkeley, CA.
Clarence Glacken's monumental survey of the history of Nature and Culture in Western thought from ancient times to the end of the eighteenth century has become a classic. No subsequent cultural history of Nature can fail to acknowledge the ideas introduced by Glacken; and no environmental history can ignore them.

Golinski, J. (2007) **British weather and the climate of enlightenment.** University of Chicago Press.
Written as a cultural history of science, Jan Golinski's book shows that the way the British related to their weather was influenced by the intellectual ideas of the Enlightenment and the changing practices of science and changing social order during the eighteenth century. In particular, a new sense of national climate emerged as weather was first measured and then analysed in new quantitative ways.

Huntington, E. (2001) **Civilisation and climate** (reprinted from the 1915 edition). University Press of the Pacific: Honolulu, HI.
This book is a classic of its time, a systematic examination of the evidence for the climate determinist outlook. Ellsworth Huntington draws together empirical evidence and an ideological reading of empire and history to paint a picture in which individual, racial and social characteristics are heavily shaped, if not controlled, by climatic factors.

Lamb, H. H. (1982) **Climate, history and the modern world.** Methuen: London.
British climatologist Hubert Lamb gathers much of his life's work and thinking into this book, in which he tells the story of why climate change on human time-scales matters. He moves from the earliest civilisations through to the twentieth century, exploring the changing relationship between physical climates and human societies.

Meyer, W. B. (2000) **Americans and their weather.** Oxford University Press.
William Meyer is a cultural geographer who surveys the multiple ways in which the American people from the seventeenth century onwards have made sense of their weather. He shows that what weather means to Americans has always been a function of culture, technology, economics and even politics.

Strauss, S. and Orlove, B. (eds) (2003) **Weather, climate, culture.** Berg: Oxford.
An edited book comprising fifteen chapters in which anthropologists, historians and geographers explore the different ways in which societies in history and around the world have related to aspects of weather and climate. Examples are included from Brazil, England, Iceland, Switzerland and Tanzania.

TWO

The Discovery of Climate Change

2.1 Introduction

In the spring of 1845, Sir John Franklin set sail from England on a naval expedition charged by the British Government with identifying and mapping an ocean channel through the Northwest Passage. This fabled trade route, linking the Atlantic and Pacific Oceans through the sea-ice of the Canadian archipelago, had been central to much of Britain's maritime exploration of the North American continent since as far back as the sixteenth century, but no-one had yet been successful in opening the Passage. Neither was Franklin. He and all his men perished in their search, prompting an upwelling of national mourning, and adding another chapter to the mythology of heroic British explorers.

Over 160 years later, changes in Arctic climate are doing what hundreds of years of exploration could not achieve – opening a clear sea passage between the two oceans. In August 2007, the Norwegian Polar Institute announced that the Northwest Passage – the sea route – was open to routine traffic, and several ships navigated the passage before it was again closed by the growing winter sea-ice. Canada has instigated new naval patrols in the region to demonstrate her territorial claims and is contemplating a new military base in the Arctic. Russia has staked its claim to 460,000 square miles of resource-rich Arctic waters by planting its flag on the North Pole's sea floor. And the prospect of new shipping routes connecting northern Canada with eastern Asia is offering new economic opportunities in the Canadian north. The small settlement of Churchill on Hudson Bay could find itself at the centre of a new network of international shipping routes.[1] The melting Arctic is just one example – a dramatic and visual one – of how changes in physical climate alter both the natural and human worlds: changing the geometry of land and ice and the tone of our political discourses, but also changing our imagination by making us rethink the significance of our narratives of the past.

It may have been unthinkable to the commissioners of Britain's nineteenth-century maritime explorations that changes in climate could bring about their long-desired goal of extending the reach of the country's imperial trading power. Yet we now believe – we now know – that, far from being constant, the essential character of physical climate is its inconstancy. The great Ice Ages in the Earth's past, only dimly apprehended by the explorers of 1845, are now an established part of our natural history. The large-scale modification of the land surface wrought by humans over long centuries, or short decades, alters the climatic properties of regions, and now a new agent of climate change – humanity itself – seems to be at work, evidenced through shrinking Arctic sea-ice, retreating Himalayan glaciers, and rising sea levels.

This chapter tells the story of how we 'discovered' that physical climates could change and, more importantly, how we 'discovered' that humans have become an active agent in these changes. Section 2.2 constructs a simple genealogy of the idea of climate change, a crucial part of which was the revolutionary proposition of the early nineteenth

---

[1] 'Canada's climate change boomtown', see http://news.bbc.co.uk/1/hi/business/7155494.stm [accessed 28 May 2008] extracted from the radio programme 'The great game in a cold climate' broadcast on BBC Radio 4 on 1 January 2008.

century that huge planetary-scale changes in climate had occurred in the distant past (Section 2.3). The bulk of the chapter (Section 2.4) then traces the growing realisation through the twentieth century that the collective impact on the atmosphere of human actions was also sufficient to change the way the climate system worked; to change the nature of weather not just locally, but globally. Rather than going into great detail here – the story has been told in much greater depth elsewhere[2] – we focus on six individuals and the contribution to knowledge they made and, with the benefit of hindsight, the significance of these contributions. These six scientists are: John Tyndall, Svante Arrhenius, Guy Callendar, Charles Keeling, Syukoro Manabe and Wally Broecker. The cultural contexts in which their science was practised becomes as important a part of the climate change story as their substantive contribution to new ways of understanding climate. The chapter concludes (Section 2.5) by reflecting briefly on how – and why – the idea of anthropogenic climate change has currently achieved such prominence in scientific, political and popular discourse.

2.2 The Genealogy of Climate Change

In Chapter 1: *The Social Meanings of Climate*, we traced the etymology of the word 'climate' back to Classical Greek culture. Cultural discourses around climate *change* also have a history, a genealogy that can again be followed back to the Greeks. Aristotle's student Theophrastus, in the third century BC, first observed and documented local changes in climate induced by human agency: the draining of marshes cooled the climate around Thessaly in Greece, while the clearing of forests around Philippi warmed the climate.[3] Not only

---

[2] See, for example, Fleming, J. R. (2005) *Historical perspectives on climate change.* Oxford University Press; and Weart, S. R. (2003) *The discovery of global warming.* Harvard University Press: Cambridge, MA.

[3] Glacken, C. (1967) *Traces on a Rhodian shore: nature and culture in Western thought from ancient times to the end of the eighteenth century.* University of California Press: Berkeley, CA, pp. 129–30.

could climates change, but human interventions in the natural world could act as the agent of change. A later Greek discourse of climate change was also constructed around changes in climate, but changes experienced through mobility – for example, the widely held view of the Classical era that Mediterranean travellers would turn black at the Equator, or else die. The experience of encountering forbidding climates through journeys into unexplored territories, and the anxieties of such climatic encounters, is a human idea that has resurfaced many times since these fearful tales were first told.

Enlightenment discourses about climate change from the seventeenth century onwards frequently concentrated around the effects of deforestation. The eighteenth-century historian Edward Gibbon could see the beneficial warming effects of tree-clearing both in changes of climate through time – the 'improvement' in European climate since Classical times, he claimed, was due to the clearing of 'immense woods'; and in the differences in climate caused by geography – the contemporary forests in Canada, he believed, subjected that land to a climate as fierce as that of ancient Germany.

Other projects were also afoot through which Enlightenment thinkers began to believe that regional climates could be subject to the will of Man. The eighteenth-century French philosopher Comte de Buffon reflected the growing ideology of human mastery when he remarked that, 'the addition or removal of a single forest in a country [by Man] is sufficient to change its temperature ... [so] ... modifying the influences of the climate he lives under ... to the point that it suits him'.[4] This philosophising was put to the test in the New World as early American colonists cleared large swathes of forest along the eastern seaboard. Hugh Williamson, an American doctor, was therefore able to remark in 1760, in front of his American Philosophical

---

[4] pp. 236–244 in Buffon (1778) *Les époques de la Nature*. Paris; cited in Boia, L. (2005) *The weather in the imagination* (trans. R. Leverdier). Reaktion Books: London, p. 65.

Society colleagues, that the climate had changed within the last forty to fifty years; the winters being less harsh, the summers cooler.[5]

Climates elsewhere in the world were also under suspicion of changing and deforestation was again to blame, but this time introduced by European settlers in the tropics. This tropical manifestation of climate change had different effects compared with its temperate cousin, since destruction of tropical forests was believed here to exacerbate the droughts that many early colonists in the eighteenth century found endemic in sub-tropical regimes. According to this view, climate change threatened not only the economics and well-being of a colony, but posed hazards to the integrity and health of the settler populations in plantation colonies.[6]

By the middle of the nineteenth century, the combined work of geologists and physicists had revealed a new dimension to climate change completely unrelated to the activities of Man: the huge swings in global climate implied by massive and ancient glaciations, a story that is told in more detail in the next section. Alongside these emerging radical ideas of climatic instability occurring over previously unimagined time-scales, observers were still grappling with the extent to which human activities could alter contemporary regional climates. Such changes appeared to be trivial when set against the large changes in climate induced by the newly found Ice Ages, and increasing volumes of meteorological data were beginning to cast doubt on the extent of substantial human influence on climate via forest clearing.

Evidence for recent changes in climate was, nevertheless, offered by some. For example, Monsieur Arago of the French Institute in Paris wrote in 1834 that, based on the evidence of grape-ripening dates, the summers in several parts of France were 'colder than they had been formerly',[7] a line of reasoning he also applied to England and her

---

[5] Cited in Glacken, *Traces on a Rhodian shore*, p. 659.

[6] p. 8 in Grove, R. H. (1998) *Ecology, climate and empire: the Indian legacy in global environmental history, 1400–1940*. Oxford University Press: Delhi.

[7] *The Times*, 18 February 1834.

apparent decline in viticulture since the days of the 'old chronicles'. Eduard Brückner, a prominent Austrian geographer, was another of this small band of climate change believers in the late nineteenth/early twentieth century, who maintained that statistical evidence *could* be found for contemporaneous changes in regional climates. Brückner demonstrated that average temperature and precipitation for areas of central Europe and Russia when measured over successive thirty-five-year periods differed substantially, claiming that such changes in climate would have implications for rivers, lakes and agriculture. His insistence that such changes were real and were related to deforestation led him to advocate in the Prussian House of Representatives what was probably the first climate change policy: a proposed law to preserve the forests of Prussia to protect the rainfall and river levels of the state.[8]

Despite the work of Brückner and a few others, the dominant mindset during the first half of the twentieth century was that physical climate was basically constant on time-scales that mattered to human thought and planning. In the context of the ever-lengthening appreciation of the deep past, however, beliefs about climate change had changed radically.

2.3 Natural Climate Change

A crucial part of the story of climate change was the dawning realisation of the multiple glaciations which the Earth has experienced over its history. That the world's climate had changed sufficiently to induce such massive glacial advances and retreats was an idea that did not emerge until the early decades of the nineteenth century. The prevailing view at the time was that of a young Earth and a stable climate,

---

[8] Brückner, E. (1890) *Klimaschwankungen seit 1700 nebst Bemerkungen über die Klimaschwankungen der Diluvialziet.* Wien and Olmütz; quoted in von Storch, H. and Stehr, N. (2006) Anthropogenic climate change: a reason for concern since the 18th century and earlier. *Geografika Annaler* 88A(2), 107–13.

punctuated by occasional natural catastrophes – earthquakes, floods, volcanic eruptions – which between them could account for all of the geological and fossil evidence of a dynamic Earth that was then being discovered. While eighteenth-century natural philosophers were constructing the notion that Man might alter regional climates through the force of his own development and technology, no-one was contemplating powerful natural forces that could radically destabilise global climate.

Not until Jean Louis Rodolphe Agassiz in the 1830s. The evidence that persuaded this 30-year-old Swiss naturalist that the world's climate *could* change in such substantial ways was presented to him and his colleagues by Alpine glaciers. A small number of Swiss scientists and engineers, partly concerned with the hazards of bursting ice-dammed lakes, had become convinced that glaciers had once extended well beyond their then known limits. It was Agassiz's older colleagues Ignace Venetz and Jean de Charpentier who convinced the ambitious young naturalist of the plausibility of such a hypothesis, but it was Agassiz alone who in July 1837 first presented a coherent Ice Age theory to the Swiss Society of Natural Sciences meeting in Neuchâtel. The idea of Ice Age climates affecting an ancient Earth became even more firmly associated with Agassiz after he subsequently published his book *Etudes sur les Glaciers* in 1840.

The idea that the Earth's climate was susceptible to such large changes in climate, and over such long time-scales, was not an easy one for mid nineteenth-century Europe to accept. It sat uneasily alongside traditional Biblical interpretations of a (fixed) created world and of the supposed beneficent hand of God preserving climatic conditions suitable for human habitation. Yet Agassiz began to find scientific converts, first in Europe and then in the United States of America after he moved there in the 1840s. By the 1860s, Agassiz's glacial theory of climatic change was firmly established on both sides of the Atlantic. At this early stage of understanding it was the glacial period(s) within the past few hundred thousand years that was under consideration; the existence of even more ancient Ice Ages remained as yet unsuspected.

The question of causation, however, remained very much in dispute. It was obvious that the Ice Age was unrelated to human influence on climate, and various naturalistic causes were proposed to displace lingering ideas of divine manipulation of the climate system. Sunspot activity, volcanic eruptions, atmospheric dust, movements in the Earth's crust and changes in the orbital characteristics of the Earth were all put forward as candidates. James Croll's astronomical theory of Ice Ages began to find favour following the publication of his 1875 book *Climate and Time*, but alternative hypotheses continued to circulate. The idea that reductions in atmospheric carbon dioxide could induce such a climatic change was also considered by John Tyndall and others (see section on John Tyndall and greenhouse gases), and prompted Svante Arrhenius to undertake the first calculations of the sensitivity of world temperature to carbon dioxide in the 1890s (see section on Svante August Arrhenius and climate sensitivity). Croll's astronomical theory found its eventual champion in the Serbian mathematician Milutin Milanković, whose 1924 elaboration of Croll's ideas thereafter became canonical as the explanation for glacial climates.

It was within this nineteenth-century world of changing views on the stability of climates that the early scientific foundations were laid for what has now become the anthropogenic theory of climate change.

2.4 Anthropogenic Climate Change

The pedigree of our scientific understanding of how humans are today altering global climate can be traced back to 1824, when the French physicist Jean-Baptiste Joseph Fourier presented an essay to the Académie Royale des Sciences in Paris on the regulation of planetary temperatures. In this essay, published later that same year, Fourier correctly understood that the atmosphere is asymmetrical with respect to the transmission of incoming solar energy and outgoing terrestrial energy – the constituent gases of the atmosphere are more opaque to outgoing thermal energy than they are to incoming short-wave solar energy. This phenomenon was later christened the 'greenhouse effect'. The story of the last 180 years of scientific thought, research and scholarship with respect to this idea is too long and convoluted to recount here. In laying the ground for later chapters, however, this scientific journey is illuminated by telling the story of six individual scientists whose contributions to the overall body of scientific knowledge about anthropogenic climate change are seminal. Each vignette describes the contributions to knowledge that were made, the context in which they were made and in which they were received, and a comment on the contemporary significance of the work of these individuals. In telling these stories, we see at work the peculiarities of science: how context, funding, patronage, technology and personality each add texture and serendipity to the production of knowledge.

1859: John Tyndall and greenhouse gases

On Wednesday 18 May 1859, after a full day's work in the basement laboratory at the Royal Institution in central London, the 38-year-old Irish scientist John Tyndall wrote in his diary, 'Experimented all day; the subject is completely in my hands!'[9] Tyndall had been experimenting with the absorptive properties of gases with a view to testing the idea that different gases, commonly found in the atmosphere, absorbed differing amounts of short-wave and long-wave radiation. This was the idea that Joseph Fourier had articulated thirty years earlier and, if true, could help to explain how the temperature of the planet was regulated. Tyndall's experiments in May 1859, which he further advanced for several years afterwards, established for the first time that the molecules of water vapour, carbon dioxide, nitrous

---

[9] pp. 36–55 in: Tyndall, J. (1855–1872) *Journals of John Tyndall Vol. 3, 1855–1872* Unpublished Tyndall Collection, Royal Institution, London.

oxide, methane and ozone each exhibit unique absorptive properties when radiant (infra-red) heat is passed through them; thus, for example, water vapour is 16,000 times more effective at absorbing infra-red radiation, molecule-for-molecule, than is oxygen.

John Tyndall's discoveries became central for understanding the heat budget of the atmosphere. By demonstrating that this group of gases – later collectively named 'greenhouse gases' – possessed distinctive radiative properties, his work opened up the possibility that, by altering the concentrations of these gases in the atmosphere, human activities could alter the temperature regulation of the planet. Although Tyndall could foresee such a possibility in the 1860s – changes in the amount of any of the radiatively active constituents of the atmosphere 'could have produced all the mutations of climate which the researches of geologists reveal'[10] – he was a long way from developing a coherent account of how human actions could induce significant changes in climate on the relevant time-scales.

Tyndall's research was published at the beginning of the intellectually tumultuous decade of the 1860s. On Thursday 24 November 1859, just six months after Tyndall's initial experimental results, Charles Darwin's book *On the Origin of Species* was published in London. Along with this challenge to the prevailing orthodoxy of a fixed biological creation, scientists were also grappling with the equally revolutionary implications of Louis Agassiz's new 'Ice Age theory' (see Section 2.2). Trying to understand the causes of these great Ice Age fluctuations in climate was one of the issues of the time. The ideas of Darwin and Agassiz were assaulting fundamental conceptions of time and stability in, respectively, biological and climatic history. John Tyndall was intimately connected with these debates. He became a close friend of Thomas Huxley and other scientists in Darwin's circle and he was consulted by Charles Lyell – who was trying to evaluate Croll's newly published orbital theory – about

[10] pp. 276–277 in Tyndall, J. (1861) On the absorption and radiation of heat by gases and vapours. *Philosophical Magazine* (Ser. 4) 22.

whether Tyndall's new radiative theory of climatic change could help unravel the causal mystery of the Ice Ages. On 1 June 1866, Tyndall replied to Lyell saying that changes in radiative properties alone were unlikely to be the root causes of glacial epochs, thus contradicting his earlier supposition of 1861 quoted above. These exchanges about theories of climatic change presaged much later arguments about the interplay between natural and human factors in the modification of global climate.

John Tyndall was one of the outstanding scientific personalities of the Victorian age and a passionate defender of the scientific naturalism which flowered in later Victorian Britain. He had a typically eclectic education, moving from school into railway surveying during the railway boom years of the 1840s, before teaching school mathematics and earning his doctorate in experimental chemistry at Marburg in Germany. His professional career was finally established when Michael Faraday invited him, in 1853, to give a series of discourses at the Royal Institution in London. Under Faraday's patronage, Tyndall was able to develop and display his talent for experimental science across an astonishing variety of subjects.

Tyndall is deservedly credited with establishing the experimental basis for Fourier's 'greenhouse effect'. Tyndall was also correct in identifying the fundamental role of water vapour in atmospheric dynamics which, he claimed, 'must form one of the chief foundation-stones of the science of meteorology'.[11] Now, about 150 years later, the differential absorptive properties of the suite of greenhouse gases he investigated – now expanded to include a range of artificial gases unknown to Tyndall, the halocarbons – still remain central to the idea of anthropogenic climate change. Subsequent work has established the global warming potentials of each of these gases with some level of precision, calculations that are pivotal in efforts to quantify the extent of human influence on the world's temperature.

[11] p. 54 in Tyndall, J. (1863) On the passage of radiant heat through dry and humid air. *Philosophical Magazine* (Ser. 4) 26.

1896: Svante August Arrhenius and climate sensitivity

By the 1890s there was widespread agreement about the nature and extent of previous glaciations on Earth (see Section 2.2), but significant disagreement about the cause of these large climatic oscillations. Croll's hypothesis of astronomical variations 'enjoyed a certain favour with English geologists',[12] but others preferred the idea that it was 'changes in the trapping of heat by vapours and gases in the atmosphere'[13] which controlled the Earth's climatic history. No-one, however, had performed the calculations to show just how sensitive the Earth's climate was to changes in atmospheric composition; a concept that has now become known as 'climate sensitivity'.

The first calculations of this parameter were performed by a 36-year-old Swedish physicist, Svante August Arrhenius, and his results were presented before the Stockholm Physical Society in 1895. Arrhenius was familiar with the conceptual 'greenhouse' thinking of Fourier, the experimental results of Tyndall from the 1860s and his later, modified, calculations of the absorption coefficients for water vapour and carbon dioxide. In a classic exercise of scientific synthesis, Arrhenius drew these ideas and data together to show that halving or doubling the concentration of carbon dioxide in the global atmosphere would lead to changes in average surface air temperature of the Earth of between about 4° and 5°C.

Arrhenius's calculations were performed by hand, not computer, and communicated by written and spoken word, not by the internet. Yet his network of scientific correspondents was considerable and his calculations – which showed that changes in atmospheric composition *could* in principle lead to glacial climates – were well

---

[12] p. 274 in Arrhenius, S. (1896) On the influence of carbonic acid in the air upon the temperature of the ground. *London, Edinburgh and Dublin Philosophical Magazine and Journal of Science* 41, 237–76.

[13] Manson, M., cited p. 15 in Mudge, F. B. (1997) The development of the greenhouse theory of global climate change from Victorian times. *Weather* 52, 13–17.

received by Italian, Swedish and American scientists. The continuing resistance to Croll's astronomical theory of Ice Ages meant that, by the turn of the century, 'the theory of global climate change [caused] by changes in atmospheric carbon dioxide levels was widely accepted'.[14]

But what caused the rise and fall of atmospheric carbon dioxide in the distant past? And could changes in atmospheric concentrations of this gas also be a contemporary phenomenon? Arrhenius was rather less interested in these questions and left it to his geologist colleagues, Arvid Högbom and Thomas Chamberlin, to argue over the carbon cycle. Although it was known that nearly a billion tons of coal worldwide was being combusted annually at the turn of the century, the belief was that little of the released carbon dioxide would remain in the atmosphere. Arrhenius, when pushed, conceded that such combustion might induce 'a noticeable increase' in atmospheric carbon dioxide over the course of a few centuries. Such warming, he deduced, might be beneficial for humanity by staving off the next glacial cycle.

Svante Arrhenius is a significant figure in the story of climate change. As so often with new scientific insights, there were things he got right and things he got wrong. He was well ahead of his time in making the first estimates of climate sensitivity – a temperature value which describes how much the world would eventually warm if the concentration of greenhouse gases in the atmosphere were doubled. Over a hundred years later and the judgement of the world's scientists is that its value probably lies somewhere between 2° and 4.5°C, although slightly lower or much higher values cannot be ruled out. The laborious manual calculations made by Arrhenius were not significantly in error, even though he couldn't have known how problematic determining this quantity was going to prove for later scientists. On the other hand, the cycling of carbon through the atmosphere, biosphere and oceans

[14] *Ibid.*, p. 15.

was poorly known in 1900, and Arrhenius grossly underestimated the rate at which carbon dioxide could accumulate in the atmosphere. Rather than taking several centuries to effect a noticeable increase, the concentration of this greenhouse gas has increased by about 40 per cent in not much more than a century, and may well have doubled before the present century is seen out.

Further understanding of the mechanisms of climate change did not proceed smoothly from Arrhenius' insights. As the twentieth century developed, the carbon dioxide theory of Ice Ages and of climate change largely disappeared from scientific discourse. The Swedish physicist Knut Ångström had suggested that the absorption power of carbon dioxide was limited by the overlapping spectral bands of water vapour, there was little agreement among geologists about what could cause carbon dioxide concentrations in the atmosphere to rise and fall in cyclical fashion, and the calculations by Milanković in the 1920s gave new credibility to Croll's astronomical theory.

But the world's temperature was now rising and a few meteorologists had begun to notice.

1938: Guy Stewart Callendar and global temperature

Many of the winters during the period from the 1910s to the 1930s were generally mild over Europe and North America, the two regions in the world where meteorological observations had their longest historical reach. These formal records lent themselves to statistical analysis, enabling the popular perception of a warming climate to be placed into a longer-term perspective. James Kincer of the US Weather Bureau had proposed in the mid-1930s that the data suggested that an 'apparent longer-term change' towards a warmer climate was taking place,[15] but he used his analysis primarily to corroborate intuitive

---

[15] Kincer, J. B. (1933) Is our climate changing? A study of long-time temperature trends. *Monthly Weather Review* 61(9), 251–9.

statements commonly being heard from the elderly that 'winters were colder and the snows deeper when I was a youngster'.

It took an idiosyncratic British mechanical engineer to put together a coherent argument as to *why* the regions around the North Atlantic were warming. Guy Stewart Callendar had an enquiring mind, trained under the influence of his father, a physics professor in London. Although his professional duties as a steam engineer at Imperial College required him to contribute to the international efforts to standardise steam tables, as a personal pursuit Callendar had become interested in Arrhenius's carbon dioxide theory of climate change, which in the 1930s lay largely dormant. Perhaps prompted by his own perceptions of mild British winters he began collecting long series of meteorological observations from stations around the world. He also made his own set of calculations about the net absorption effect of carbon dioxide in the atmosphere, following in Arrhenius's footsteps but using more recent coefficients. His hunch was simple: 'As man is now changing the composition of the atmosphere at a rate which must be very exceptional on the geological time-scale, it is natural to seek for the probable effects of such a change.'[16] Callendar believed these effects to be a warming climate which could be detected in the observations.

Callendar completed his work during the mild English winter of 1936/7. Although no meteorologist, he first presented his findings to a London meeting of the Royal Meteorological Society in February 1938, a few weeks before Hitler's annexation of Austria. With hindsight, this study can be seen as the first attempt at detecting and attributing large-scale climate change to human-induced emissions of greenhouse gases (see Box 2.1).

---

[16] p. 38 in Callendar, G. S. (1939) The composition of the atmosphere through the ages. *Meteorological Magazine* 74, 33–9.

Box 2.1: Detecting and Attributing Anthropogenic Climate Change

The paper that Guy Stewart Callendar read to the Royal Meteorological Society on Wednesday 16 February 1938 was the first systematic attempt to link together the three pillars of the idea of anthropogenic climate change: the physical theory of carbon dioxide and the greenhouse effect, the rising concentration of carbon dioxide in the atmosphere, and the increase in world temperature. His was the first study of what has now become known as the detection and attribution of anthropogenic climate change. We can compare Callendar's pioneering effort with the more recent considered assessments about detection and attribution contained in the four reports of the Intergovernmental Panel on Climate Change (IPCC) published between 1990 and 2007.

Callendar calculated that the expected increase in world temperature on the basis of physical theory and his carbon dioxide concentration estimates, 'to be at the rate of 0.3°C per century at the present time'. He followed this by stating baldly that 'world temperatures have actually increased at an average rate of 0.5°C per century during the past half-century.' [17] Although Callendar was clear in his own mind that he had found strong evidence for anthropogenic global warming, many of his listeners that day in London were sceptical. Sir George Simpson, then director of the British Meteorological Office, thought that 'his results must be taken as rather a coincidence'.[18] The five other discussants in the Reading Room of the Society that day also expressed considerable scepticism over Callendar's claims.

---

[17] p. 223 in Callendar, G. S. (1938) The artificial production of carbon dioxide and its influence on temperature. *Quarterly Journal of the Royal Meteorological Society* 64(2), 223–40.

[18] *Ibid.*, p. 237.

Just over fifty years later, in September 1990, the First Assessment Report of the IPCC was announced in front of a rather larger audience of the world's media gathered together in Bracknell near the headquarters of the UK Meteorological Office. This report, published under the auspices of the United Nations, contained a considered statement about the question of detection and attribution: 'Our judgement is that global-mean surface air temperature has increased by between 0.3° and 0.6°C over the last hundred years ... the size of this warming is broadly consistent with predictions of climate models, but it is also of the same magnitude as natural climate variability.' This guarded conclusion was more cautious than Callendar's in 1938, yet it initiated what was to become the central scientific question about climate change over the next two decades: the detection and attribution of anthropogenic climate change.

Successive reports from the IPCC reveal the incremental way in which these scientific assessments extended the belief that formal detection and attribution of human influence on global climate had been attained. Thus, in 1996, the Second Assessment Report stated that 'the balance of evidence suggests that there is a discernible human influence on global climate', while the Third Assessment Report in 2001 concluded that 'most of the observed warming over the last fifty years is likely[19] to have been due to the increase in greenhouse gas emissions'. In the Fourth Assessment Report, published in February 2007, the statement had hardened into a more bullish: 'Most of the observed increase in global average temperatures since the mid twentieth century is very likely[20] due to the observed increase in anthropogenic greenhouse gas concentrations.'

---

[19] According to IPCC conventions, 'likely' implies a chance of between 66 and 90 per cent that the statement is correct.

[20] According to IPCC conventions, 'very likely' implies a greater than 90 per cent chance of the statement being correct.

These five statements, spread over seventy years of scientific endeavour, reveal two important things about science. They reflect a conventionally assumed attribute of science as being a cumulative enterprise, in which knowledge is shaped and influenced by previous claims and propositions, often building and consolidating confidence in what is already known. Thus one follows the transition from ignorance (before 1938), through informed speculation (Callendar in 1938), to increasing levels of confidence (IPCC between 1990 and 2007) about the detection of human influence on world climate. But these sequential statements challenge another attribute commonly associated with science – an adjudicator of objective and certain truth. These statements are constituted in the language of belief and probability. At each stage of the process, substantial judgement on the part of the scientist(s) is required; a judgement formed through dialogue, dispute and compromise rather than through detached and disinterested truth-seeking. Knowledge about climate change is being communicated in conditional terms. This is a theme to which we will return in Chapter 3: *The Performance of Science*.

Callendar's work was not well received. Apart from scepticism from the meteorological professionals, there were others who questioned his estimates of carbon dioxide in the atmosphere and his absorption calculations. His cause was not helped by the perception, and later quantification through the 1950s and 1960s, that temperatures in Europe and North America had begun to fall. Callendar himself had proposed a verification of his ideas by stating at the end of his 1938 paper that 'the course of world temperatures during the next twenty years should afford valuable evidence as to the accuracy of the calculated effect of atmospheric carbon dioxide'.[21] At face value, and on his own terms, the evidence seemed to contradict his theory.

[21] Callendar, The artificial production of carbon dioxide, p. 236

Many years later, in November 1960, Callendar wrote in his personal papers of the continuing unpopularity of the carbon dioxide theory of climate change, citing four reasons why people didn't accept it:

- The idea of a single factor causing worldwide climatic change seems impossible to those familiar with the vast complex of forces on which any and every climate depends.
- The idea that man's actions could influence so vast a complex is very repugnant to some.
- The meteorological authorities of the past have pronounced against it, mainly on the basis of faulty observation of water vapour absorption.
- Last, but not least, they did not think of it themselves![22]

Callendar died in 1964, long before his 1938 ideas became common scientific currency. It was another twenty-two years after his death before the first truly global – land and marine – temperature series was published, and another forty-three years before the IPCC announced what Callendar had argued in 1938: that the rise in world temperature was very likely to be a result of the rise in carbon dioxide. A deeper irony was that, whereas Callendar's physical reasoning from the 1930s has proved broadly correct, his interpretation of the *significance* of what he claimed is not the way in which climate change is now viewed. 'The combustion of fossil fuel … is likely to prove beneficial to mankind in several ways … for instance, the increases in mean temperature would be important at the northern margins of cultivation, and the growth of favourably situated plants is directly proportional to the carbon dioxide pressure. In any case, the return of the deadly glaciers should be delayed indefinitely.'[23]

[22] Guy Stewart Callendar papers, CRU/CAL/L/LE2a, University of East Anglia, Norwich, UK.
[23] Callendar, The artificial production of carbon dioxide, p. 236

1957: Charles David Keeling and the carbon cycle

Callendar's pioneering research efforts in the late 1930s had required no external funding; he undertook his calculations in the evenings and weekends while working for Imperial College. For many scientific innovations, however, funding the research to test ideas can be more difficult than conceiving the ideas in the first place. This is certainly true in the case of large-scale and continuous field measurements.

By the mid-1950s there were a small number of geophysicists who clearly understood that, without knowing the fate of carbon dioxide molecules emitted into the atmosphere, the arguments over Arrhenius's and Callendar's ideas could not be settled. How much of the carbon dioxide emitted from fossil fuel combustion could the oceans absorb? How fast was carbon accumulating in the atmosphere? No accurate or universal measurements of carbon dioxide concentrations existed, and there were certainly no standardised historical data with which to monitor change over time. Colleagues Roger Revelle at Scripps Institution of Oceanography in California and Hans Suess at the University of Chicago saw a great opportunity on the back of the 1957/8 International Geophysical Year (IGY) to get the funding needed to set up new monitoring sites. The IGY was a major international scientific collaboration aimed at quantifying the state of the global environment, while doubling up for the USA and the USSR as a convenient front for extending the reach of military surveillance and as an exercise in strategic geopolitical posturing.

Revelle and Suess, however, and the young post-doctoral scientist they hired to run these new carbon dioxide measuring sites – Charles David Keeling – were simply grateful to acquire the funding to pursue their scientific goal: 'To establish a reliable "baseline" carbon dioxide level which could be checked ten or twenty years later.' [24] Keeling established two sites at which he initiated a routine series of

[24] Keeling, C.D. cited p. 439 in Weart, S.R. (2007) Money for Keeling: monitoring CO_2 levels. *Historical Studies in the Physical and Biological Sciences* 37(2), 435–52.

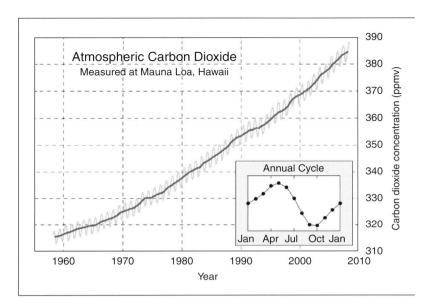

FIGURE 2.1: *Measurements of atmospheric carbon dioxide concentration (ppmv) at Mauna Loa Observatory, Hawaii. The concentration has risen by about 25 per cent since Charles Keeling started these measurements in 1957/8. Source: NOAA/Wikipedia.*

measurements using a new infra-red analyser technique which he had perfected: one site was at the Mauna Loa Observatory in Hawaii, on top of the world's largest volcano, and the other was at the American scientific base at the South Pole. He started these measurements of atmospheric carbon dioxide concentrations in September 1957 at the South Pole and in March 1958 at Mauna Loa – measurements that are still made routinely today, more than fifty years later (see Figure 2.1).

Keeling's measurements were more successful than either he or Revelle had anticipated. Within eighteen months of starting, Keeling had revealed that the carbon dioxide concentration was rising at both sites, by between 0.5 and 1.3 parts per million (ppm) per year. This was the first incontrovertible evidence for the contemporary increase of this greenhouse gas in the atmosphere that Callendar had estimated and upon which Arrhenius had earlier speculated. Keeling offered caveats about his findings, wondering, for example, about the varying roles of oceans and land biota in modulating these increases year by year; yet these pioneering measurements laid the foundation for all subsequent work on understanding the carbon cycle and the human role in modifying it.

But the importance of these measurements was realised at the time by only a few. Throughout the 1960s, Revelle, Suess and Keeling, and then later in the 1970s Keeling alone, were engaged in frequent negotiations to secure funding to keep the routine measurements going. They ceased altogether for several months in the first half of 1964 as US Weather Bureau support was withdrawn, and even as late as 1979, officials at the US National Science Foundation told Keeling that they would not support routine monitoring indefinitely. Throughout these decades, and right up until his death in June 2005, Keeling maintained a belief in the importance of the measurements he was taking or supervising, well before their significance became apparent to others, or even fully to himself.

1975: Syukuro Manabe and climate models

Keeling's carbon dioxide curve from Mauna Loa continued rising through the 1960s and 1970s, from 314 ppm in 1957 to 331 ppm in 1975. Although Northern Hemisphere land temperatures were not keeping track with this increase, new developments in atmospheric science all pointed towards Callendar's (1938) proposition being correct. The question Arrhenius had provided one answer to in 1896 became increasingly important: exactly how sensitive was the Earth's climate to a doubling of atmospheric carbon dioxide? Calculations made using one-dimensional radiative–convective models in the 1960s and early 1970s had suggested that climate sensitivity was in the range 1.9°–10°C or more. This wide range of published values reflected different modelling approaches and, in particular, different representations of the feedback processes associated with an enhanced greenhouse effect. The role of water vapour, the importance of which Tyndall had recognised back in the 1860s, was particularly problematic. In 1967, atmospheric scientists Syukuro Manabe and Richard Wetherald at Princeton University, New Jersey, had undertaken one such calculation that suggested a climate sensitivity of 2.3°C, but there clearly was no consensus about this value nor about the most appropriate modelling approach through which it could be estimated.

Developments in computer technology and in the numerical weather prediction models used for short-range weather forecasting had, by the early 1970s, spawned the development of new general circulation models of the global atmosphere with which climate could be simulated. Manabe had developed one of these models at the Geophysical Fluid Dynamics Laboratory at Princeton and, together with his colleague Wetherald, used it to conduct the first model experiment which explicitly simulated the three-dimensional response of global climate to a doubling of atmospheric carbon dioxide concentration. Such a model, used for this purpose, represented an advance over previous work in a number of ways. Heat transport by large-scale eddies in the atmosphere were explicitly represented, the water vapour feedback was simulated, and the effect of the ice–albedo feedback was quantified for the first time.

The results of this emblematic model experiment have withstood remarkably well the subsequent advancements in understanding of the physics of the climate system and the capacity to model it. Manabe's model computed a climate sensitivity of 2.9°C; higher than many of the values derived from the best radiative–convective models of earlier years, and well within the range now cited by the IPCC. Their model also allowed them to determine, for the first time, the vertical and latitudinal gradients of the air temperature response to carbon dioxide doubling: greater warming at the poles was revealed, more moderate warming in the tropics, and cooling in the stratosphere; features of the climate system response to elevated greenhouse gas concentrations that remain central to studies directed to the detection of the human fingerprint on global climate change.

Manabe and Wetherald were acutely aware of the limitations of their model and therefore of the limitations to their results.[25] The horizontal spatial resolution of their model was about 500 km (meaning that islands as large as Iceland were invisible to the model), the model had no ocean (simply a 'wet surface' covering 70 per cent of the Earth's surface with infinite water available for evaporation), and there were no interactive clouds. This latter deficiency meant that one important feedback process – cloud cover changes – could not be simulated by their model. In comparison with later Earth system models their early version was undoubtedly crude, and Manabe did not exaggerate the importance of their findings, stating that 'it is not advisable to take too seriously the quantitative aspect of the results obtained in this study'.[26]

Despite these limitations, Manabe's pioneering experiment using the emerging technology of supercomputing opened the way for many more such models to be developed in climate modelling laboratories. Such models are now widely used for exploring the effect of rising greenhouse gas concentrations on global climate. Looking back from the twenty-first century and from the perspective of our artificially warmed Earth, Manabe and Wetherald's study represents a significant achievement and a notable milestone in the scientific study of anthropogenic climate change. Yet Manabe's warning about not taking the results of such a study 'too seriously' remains to be examined. In

[25] Their paper gives a full account of the 'mistake' they made when fixing the boundary between permanent and temporary sea-ice. They mistakenly programmed the boundary to occur at −25°C rather than at −10°C, their intended value. Manabe's scrupulous adherence to the principles of scientific practice meant that since they were unable to re-run the model with the correct value – presumably because of limited computer resources – they fully acknowledged their mistake in the published paper. Indeed, they turned it to their advantage by arguing that given the similarity between satellite-derived and their modelled surface albedo at high latitudes, −25°C may anyway have been a more appropriate value to have used!

[26] p. 13 in Manabe, S. and Wetherald, R. T. (1975) The effects of doubling the CO_2 concentration on the climate of a general circulation model. *Journal of the Atmospheric Sciences* 32, 3–15.

what sense are computer models predicting reality? Are they 'truth machines', to be believed in by all, or do they act more like metaphors to help us to think about the future? The nature of climate models and their uses in policy making will be examined in Chapter 3: *The Performance of Science*.

1987: Wallace S. Broecker and abrupt climate change

In the 1830s, the surface geological evidence of past glaciations in Switzerland extending well beyond existing glacier limits convinced Louis Agassiz that climates must have changed massively in prehistoric times. About 130 years later, in the 1960s, the first cores drilled through the Greenland Ice Sheet to bed-rock began to yield evidence offering a different perspective on the Earth's climatic history. The first of these ice cores was at Camp Century, a US military air base in northwest Greenland, and extended through nearly 1.4 km of ice, capturing indictors of climatic behaviour from nearly 100,000 years of Earth history. Later ice cores followed in the 1970s and 1980s, in Antarctica as well as in Greenland, and their evidence suggested that transitions between glacial and inter-glacial climates were not necessarily smooth oscillations extending over thousands of years. The Earth's climate was perhaps capable of substantial changes on much shorter time-scales, perhaps just a few hundred years, or even less than a century.

American oceanographer Wallace S. ('Wally') Broecker was one of the most outspoken of the group of scientists who exploited this new evidence for developing a new conception of climate change. In 1987 he wrote an article in the journal *Nature* in which he connected the evidence of relatively rapid changes in climate in the past to the possibility that anthropogenic climate change in the future might also trigger abrupt changes in aspects of the Earth's climate. The article was titled 'Unpleasant surprises in the greenhouse?' and his argument was that, 'We have been lulled into complacency by model simulations that suggest a gradual warming over a period of about

100 years ... these [ice-core] records indicate that the Earth's climate responds in sharp jumps which involve large-scale re-organisations of Earth's system ... I fear that the effects of [global warming] will come largely as surprises.'[27]

Broecker drew attention, in particular, to the possibility that a large-scale feature of the ocean circulation known as the thermohaline circulation, which yields the Gulf Stream in the North Atlantic, might be one such mechanism in the Earth's system that could be destabilised by rising greenhouse gas concentrations. If the thermohaline circulation weakened substantially over a short period of time, or even collapsed entirely, then the climates of the North Atlantic Basin could be quite radically altered.

This new thinking about climate change in the late 1980s was to lead scientists to find new ways of conceiving, representing and modelling climate change. The Earth's climate system needed to be understood as a whole – atmosphere (including the stratosphere), oceans, ice sheets, forests, lakes, land cover – and the interactions between all of these components were fundamental for understanding the way that the system as a whole behaved. Humans, however, were not accommodated in these models. Ideas such as thresholds, abrupt and non-linear changes, and 'tipping points' became part of the new paradigm of Earth system science. These ideas do not, however, present themselves to us, unambiguously, through observed evidence. As with the Greek idea of *klimata*, the metaphor of Gaia for the Earth system and the idea of 'tipping points' in the climate system are ways of seeing the world; ways of believing. The Greek notion of *klimata* lasted for nearly 2,000 years. It remains to be seen how durable and powerful these new conceptions of climate change, pioneered by Broecker and like-minded colleagues, will prove to be.

[27] p. 123 in Broecker, W. S. (1987) Unpleasant surprises in the greenhouse? *Nature* 328, 123–6.

2.5 Climate Change Today

In the 160 years since explorer John Franklin perished in the Canadian Arctic, the enterprise of science has succeeded in overturning the old, yet persistent, Greek idea of climate as a stable property of the natural world. 'Climate is what you expect, weather is what you get' is no longer an adequate aphorism because we have come to believe that climate, just like weather, is constantly changing. Our expectations of climate have changed. This is the insight that Louis Agassiz applied over the very longest of time-scales. This too is the more recent insight offered us by Tyndall, Callendar, Broecker and colleagues.

The view, dominant over the first half of the twentieth century, which led Hubert Lamb to remark, in 1959, 'not so very long ago ... climate was widely considered as something static, except on geological time-scale[s], and authoritative works on the climates of various regions were written without allusion to the possibility of change',[28] is as anachronistic to us today as the notion of a flat Earth. Physical climates change, they change on all time-scales, and we humans have become an active agent of change. But this alteration in perspective did not happen instantly, and it was not driven purely by science. This chapter concludes by reflecting briefly on how and why the idea of anthropogenic climate change has today achieved such prominence in scientific, political and popular discourse. We do so by considering four inflection points in recent decades – four moments in wider political or cultural history when the idea of climate change took on new significance: the 1960s environmental awakening, the 1972 Stockholm Conference on the Environment, the 'greenhouse summer' of 1988, and the singularity of the events of 11 September ('9/11') in 2001.

The early suggestions of human-induced climate change from Arrhenius in the 1900s and Callendar in the 1930s were generally

---

[28] p. 299 in Lamb, H. H. (1959) Our changing climate, past and present. *Weather* 14, 299–318.

associated with relatively benign, or even positive, consequences for society. This contrasts sharply with the tone of the current discourse about climate change, which is predominantly one of danger and catastrophe (see Chapter 7). The origins of discourses about dangerous climate change can be traced back to the environmental awakening of the early 1960s in Europe and North America, prompted in part by Rachel Carson's book *Silent Spring*, published in September 1962. One of the first associations of anthropogenic climate warming with notions of danger was in a 1963 conference of scientists convened by the Conservation Foundation of New York, which warned of a 'potentially dangerous atmospheric increase of carbon dioxide'.[29] The idea of a human-induced change in world climate found sympathy in the broader currents of intellectual thought in the 1960s and the emergence of a new environmentalism.

The first governmental and international assessments of the prospects of climate change were conducted during this period. In the USA, the President's Scientific Advisory Committee published a report on *Restoring the Quality of our Environment* in 1965, which included a specific section on climate change, and in 1971 an international *Study of Man's Impact on Climate* was published under the auspices of the Swedish Academy of Science. But it was the first United Nations Conference on the Environment in Stockholm in 1972 which was seminal in establishing more popular awareness about the potential extent of human influence on the world's environment, including its climate. The mood of concern was further heightened by the report *Limits to Growth*, published by the Club of Rome in 1972, and by the 1973 world oil crisis. This anxious mood allowed a new genre of popular climate science books to emerge during the mid-1970s – for example, Lowell Ponte's *The Cooling* (1976), Steve Schneider's *The Genesis Strategy: Climate and Global Survival* (1976) and John Gribbin's

---

[29] Cited p. 353 in Weart, S. R. (1997) Global warming, cold war and the evolution of research plans. *Historical Studies in Physical Biological Sciences*, 27(2), 319–56.

Forecasts, Famines and Freezes (1977). Although there was significant contention about whether the direction of human influence was for global warming or global cooling, the idea that humans were now a climatic force on the planet was firmly established.

A decade later, by the mid-to-late 1980s, the dominant scientific opinion had settled firmly on the prognosis of future warming. The emergence of anthropogenic global climate change as a significant public policy issue induced a heightening of anxiety. The term 'climate catastrophe' in the context of anthropogenic climate change first appeared in the German language in the cultural magazine *Der Spiegel* in April 1986 and, following the 'greenhouse summer' of 1988 (see Box 2.2), the idea of climate change penetrated more deeply into popular culture in the West, although more superficially – or not at all – in other parts of the world. The wider geopolitical resonance of climate change was linked with the collapse of the Soviet Union in 1989. Fears of Cold War destruction were displaced by those associated with climate change, prompting the observation at the time from cultural theorist Andrew Ross that, 'apocalyptic fears about widespread droughts and melting ice caps have displaced the nuclear threat as the dominant feared meteorological disaster'.[30]

Box 2.2: What Happened to Climate Change in 1988?

Unlike the hole in the ozone layer revealed by atmospheric chemist Joe Farman in 1985 or the steady rise in atmospheric carbon dioxide shown by Charles Keeling in 1960, there was no major new scientific discovery about climate change in 1988. Yet something happened in that year to bring the idea of anthropogenic climate change into the foreground of public consciousness. What happened was a

---

[30] p. 8 in Ross, A. (1991) Is global culture warming up? *Social Text* 28, 3–30.

convergence of events, politics, institutional innovations, and the intervention of prominent public and charismatic individuals.

The year started with a claim, in January, that 1987 had been the warmest year ever recorded in the 130-year global temperature series published by the University of East Anglia in the UK. The five warmest years recorded at that time were all registered in the 1980s. At the end of May, and taking inspiration from the previous year's meeting in Montreal about the protection of the ozone layer, the first major intergovernmental conference on climate change was held in Toronto with representatives from forty-eight nations. The conference statement on 'The Changing Atmosphere: Implications for Global Security' called for a 20 per cent reduction in carbon emissions from 1988 levels among the industrialised nations by 2005. And it was in 1988 that the World Meteorological Organization formally approved, at its 40th Executive Council, the establishment of a new international scientific assessment panel, to be called the Intergovernmental Panel on Climate Change (IPCC), a decision subsequently backed by the UN Environment Programme. In November the IPCC held its first working session in Geneva, leading eventually to the publication of its First Assessment Report in 1990.

Events too, contributed by the climate system, added to the momentum. As a significant drought developed across the US Midwest, a Senate hearing in Washington on Thursday 23 June took evidence from Jim Hansen, a NASA scientist from New York, in which he famously said: 'It is time to stop waffling so much and say that the evidence is pretty strong that the greenhouse effect is here.' From 1988 onwards the number of annual Congressional hearings in Washington on climate change averaged about ten per year compared with fewer than two previously (see Figure 2.2). Not only did George Bush (Senior) and Michael Dukakis take note – Bush later claiming, in the race for the White House, that he would be remembered as the 'environmental President' – but Soviet President Mikhail Gorbachev

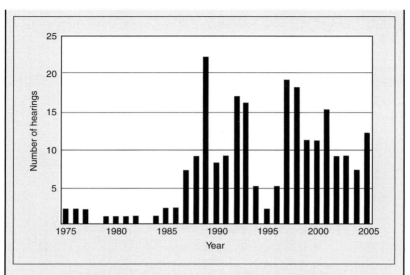

FIGURE 2.2: *Annual frequency of US Congressional hearings on climate change between 1975 and 2005.*
Source: Anon. (2006).

added a Soviet voice to the green debate for the first time. In the UK, Prime Minister Margaret Thatcher made a speech to the Royal Society in London in September which drew attention to the urgency of tackling climate change: 'It is possible that … we have unwittingly begun a massive experiment with the system of this planet itself. We need to … consider the wider implications for policy for energy production, for fuel efficiency, for reforestation.'[31]

By the end of the year *Time* magazine had picked up the mood of public concern in the Western world and offered its readers a front cover nominating 'Endangered Earth' as 'Planet of the Year', depicting the globe wrapped in polythene and tied with rope, highlighting the Earth's precarious situation and our casual treatment of it. The idea that humans not only could, but really were, warming the planet's climate had firmly embedded itself well beyond the confines

---

[31] Margaret Thatcher's speech to Royal Society, London, 27 September 1988.

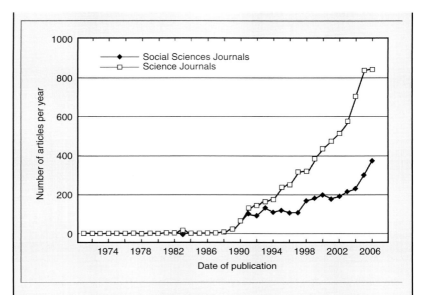

FIGURE 2.3: *Number of articles published annually in science journals (white squares) and social science journals (black diamonds) between 1970 and 2006 mentioning 'climate change' or 'global warming' (but not 'pollution') as found in the ISI Web of Knowledge. Mentions are in title, abstract or keywords.*
Source: Goodall (2008).

of science, among world leaders, national politicians and the media. For the international research community, 1988 was undoubtedly the year when climate change 'took off' (see Figure 2.3). Yet this was still a very geographically skewed penetration of the idea of climate change. Few citizens in India, China or Africa, for example, were talking about it and, in these and other nations, climate change had certainly not yet emerged as a public policy issue.

The language and metaphorical constructions of fear and catastrophe regarding climate change have been embellished substantially in the period following 9/11. The 'war on terror' provided a new benchmark against which the dangers of future climate change could be referenced, while new linguistic and metaphorical repertoires have been developed. Commenting on an extensive 2006/7 study of climate change discourses found in the British media, and in the pronouncements of senior politicians and leading public commentators, Simon Retallack of the Institute for Public Policy Research in London reflected on the significance of this language: 'The alarmist repertoire uses an inflated language, with terms such as "catastrophe", "chaos" and "havoc", and its tone is often urgent. It employs a quasi-religious register of doom, death, judgement, heaven and hell. It also uses the language of acceleration, increase, intractability, irreversibility and momentum.'[32]

At the same time, enhanced Earth system modelling capabilities have opened up new scenarios of the climatic future, simulating our alleged rapidly impending approach to triggering major alterations to large-scale functions of the Earth system, for example the melting of the Greenland Ice Sheet, the massive release of methane hydrates in the tundra, or a redirection of the thermohaline circulation of the world's oceans. These prospective futures, given virtual reality through computer modelling, have been grouped together and communicated widely using Malcolm Gladwell's 'tipping point' metaphor, further nourishing the discourse of global climate catastrophe. Not only does this discourse find saliency in the media, but also through a new cohort of popular science books – for example, Fred Pearce's *With Speed and Violence: Why scientists fear tipping points in climate change* (2007) or Mayer Hillman's *The Suicidal Planet: How to prevent global climate catastrophe* (2007).

This contemporary climate discourse of fear, constructed around looming and apocalyptic changes in future climate, resonates with discourses of the past. As we saw in Chapter 1: *The Social Meanings of Climate*, human cultures have always been capable of constructing

---

[32] p. 55 in Retallack, S. and Lawrence, T. (2007) *Positive energy: harnessing people power to prevent climate change*. Institute of Public Policy Research: London.

narratives of fear around their direct or vicarious experience of 'strange', unknown or portended climates. Yet these discourses of fear are always situated – geographically, historically and culturally. They are not imposed by Nature, they are created through Culture.

2.6 Summary

This chapter has helped us understand the places where the idea of anthropogenic climate change first emerged and when and how the idea achieved wider prominence. Some of the roles played by key individual scientists in this story, and their contributions to scientific thinking, have been revealed. But this story about the idea of climate change is not a simple one of science progressing purposefully in a straight line from blissful ignorance to a state of confident knowledge. We have shown how the idea of climate change meant different things in different places and at different times in the past, especially how understanding the causes of climate change, and the implications of these causes, has been fluid and contested. The six years in which we briefly stopped – 1859, 1896, 1938, 1957, 1975 and 1987 – and the contributions to knowledge made by the six scientists investigated, also tell us something about the contexts and contingencies that affect and shape the scientific process.

John Tyndall's work in his underground laboratory in the late 1850s captured the experimental dimension of science, the controlled conditions in which Nature can be made to perform and, as Tyndall said at the time, 'to reveal all of her secrets'. A generation later, these 'secrets' were being used by another scientist who wanted to demonstrate a particular idea: that variations in atmospheric carbon dioxide could be a credible explanation for the newly discovered Ice Ages of Earth's prehistory. Arrhenius synthesised available knowledge but applied it to the wrong problem, largely missing the significance that we now attach to his calculations of climate sensitivity. Science has not always required large investments of public money, and Guy Callendar's solitary work in the late 1930s demonstrates that an outsider to the disciplinary establishments within science – but an outsider determined to follow a hunch – could make a seminal contribution to knowledge. In 1938, he published what was, in effect, the earliest detection and attribution study of anthropogenic global climate change, a conceptual framework which has been refined and exploited, particularly over the last twenty-five years. While Callendar's 'expert judgement' was called into question in the following two decades, the process of making judgements or collective assessments of knowledge and uncertainty remains an essential foundation of knowledge claims about climate change.

When significant amounts of funding *are* required to mount large-scale field measurements, opportunism often plays a part. Exploiting opportunities afforded by the Cold War and the 1957 International Geophysical Year, Roger Revelle and Charles Keeling were able to initiate (and then sustain), routine and accurate measurements of atmospheric carbon dioxide – measurements which have become a central piece of evidence for anthropogenic climate change. Developments in technology are also integrally bound up in the enterprise of science – for example, computational power and ice-coring techniques – and these have been especially important in explaining the emergent knowledge about climate change in recent years. Syukuro Manabe, for example, exploited new supercomputing technology in the mid-1970s to pioneer what has become a hegemonic role of models in the study of climate change; while Wally Broecker, in the late 1980s, used new evidence from deep ice cores to introduce a new conceptual paradigm for how we think about climate change.

The story told thus far leaves out the contributions to discourses about climate change which have come from the social sciences and humanities; it has focused largely on the privileged position that the natural sciences have acquired in shaping this story. We will see in later chapters why the economic, ethical, psychological and political dimensions of climate change are so challenging but, given the pre-eminent position that natural science still retains in climate change debates, we first need to enquire into the nature of such knowledge. How stable is scientific knowledge about climate change? How is such natural science research funded, and how is this research received by society? Do these knowledge claims travel well across disciplinary boundaries and between cultures? We need to investigate the idea that one of the reasons we disagree about climate change is because we disagree about the nature of scientific knowledge and its role in policy making.

FURTHER READING FOR CHAPTER 2

Flannery, T. (2006) **The weather makers: our changing climate and what it means for life on Earth.** Penguin: London.
Tim Flannery's book presents a racy, populist account of the emergence of climate change and the roles of various scientists and their ideas (as well as much else about the consequences of climate change and his preferred solutions). A good easy read, as long as you recognise the polemical nature of his writing.

Fleming, J. R. (1998) **Historical perspectives on climate change.** Oxford University Press: Oxford and New York.
The finest single account yet published of the various stages in the 'discovery' of the idea of anthropogenic climate change, starting with the Enlightenment and ending in the 1970s. There is a second 2005 edition in paperback.

Fleming, J. R. (2007) **The Callendar effect: the life and work of Guy Stewart Callendar (1898–1964).** American Meteorological Society: Boston, MA.
This is the first published biography of Guy Callendar and is a very readable account of his life and work, notably his prescient contributions to the ideas behind global warming: rising world temperature, rising concentrations of carbon dioxide, and infra-red sky radiation. It is also meticulously sourced and cites all of Callendar's published works.

Imbrie, J. and Imbrie, K. P. (1979) **Ice Ages: solving the mystery.** MacMillan: London.
A good introduction to the early nineteenth-century discovery and debates around the idea of Ice Age climates, although this would need supplementing with a more recent account of how our ideas about the Ice Ages have changed in the last thirty years.

Schneider, S. H. (1989) **Global warming: are we entering the greenhouse century?** Sierra Club Books: San Francisco, CA.

Steve Schneider has had one of the longest scientific careers in the field of climate change. This book is an interesting read, since it captures how the science and policy of anthropogenic climate change were understood in the late 1980s, just at the time that serious politicians first started taking an interest in the questions raised. The book covers science, politics and the media, written by someone who was at the centre of these debates then, as indeed he remains today.

Von Storch, H. and Stehr, N. (2006) Anthropogenic climate change: a reason for concern since the 18th century and earlier. **Geografiska Annaler** 88A(2), 107–13.

A reminder, if one were needed, that climate change did not start with the IPCC and the Kyoto Protocol. This short article makes the point crystal clear by citing briefly a variety of discourses about climate change from the seventeenth century through to the 1960s. The nineteenth-century Austrian geographer Eduard Brückner is used as an illustrative character in this story.

Weart, S. R. (2003) **The discovery of global warming.** Harvard University Press: Cambridge, MA.

Another careful account of the 'discovery' of the idea of anthropogenic climate change, with greater attention paid to the American contributions of the 1950s and 1960s. This is rather less assured than the account by Fleming, but has the added advantage of being accompanied by an extensive website – www.aip.org/history/climate/ – hosted by the American Institute of Physics, which allows more interactive investigation, and also a mammoth bibliography. There is a second, expanded, 2008 edition in paperback.

THREE

The Performance of Science

3.1 Introduction

In February 2007, American physicist Fred Singer and biologist Dennis Avery published a book in the USA called *Unstoppable Global Warming: Every 1,500 years*. The authors argued that the warming currently being observed around the world is part of a natural 1,500-year cycle in solar energy and that, consequently, attempts to reduce global warming this century by controlling human-originated emissions of greenhouse gases are largely futile. That same month the Fourth Assessment Report from the Intergovernmental Panel on Climate Change (IPCC) was published. Here it was claimed that 'most of the observed increase in global average temperatures since the mid twentieth century is very likely [i.e. more than 90 per cent chance] due to the observed increase in anthropogenic greenhouse gas concentrations'.[1]

These claims cannot both be true: the observed warming cannot be largely due to changes in solar energy *and* at the same time be due mainly to rising greenhouse gas concentrations. Both claims might be

[1] p. 10 in IPCC (2007) *Climate change 2007: the physical science basis*. Contribution of Working Group I to the Fourth Assessment Report of the IPCC. Cambridge University Press.

wrong, but they cannot both be right. One common expectation of science is that it should be able to adjudicate between such competing claims to truth: seeking out evidence, testing that evidence and distinguishing between fact and error.

The above example concerns the ability of science to evaluate the veracity of the proposed causes of climate change. Another important contribution of science to debates about climate change is its attempt to establish the value of what is known as the climate sensitivity (see p. 47 for definition). For many years – since 1979 and the first US National Academy of Sciences report, *Carbon Dioxide and Climate* – numerous scientific assessments have suggested that this value is likely to lie in the range 1.5°–4.5°C. What has been harder to establish is whether the climate sensitivity is more likely to be at the low end of this range than at the high end or, indeed, what is the likelihood that the value lies outside this range altogether. The report from the IPCC in February 2007 stated that the climate sensitivity 'is likely [more than 66 per cent chance] to be in the range 2° to 4.5°C … and is very unlikely [less than 10 per cent chance] to be less than 1.5°C'.[2] Here is another customary role for science: to be able to make robust statements about the likelihood of certain future physical events occurring. It matters whether the climate sensitivity is 4.5°C or more, or whether it is 1.5°C or less; and science seems the best means we have of arriving at such a judgement.

Let us consider a third example of a possible role for science in climate change negotiations. During 2007 the international aid organisation, Christian Aid, ran newspaper and magazine advertisements urging their supporters in the UK to lobby the British Government to tighten the provisions of the forthcoming Climate Change Bill. The campaigners were lobbying for the Bill to set the

[2] *Ibid.*, p. 13. A subsequent work by Roe, G. H. and Baker, M. B. (2007) Why is climate sensitivity so unpredictable? *Science* 318, 629–32, has suggested that further narrowing of this range, or offering more accurate probability assessments, is unlikely to be achieved, due to the innate complexity of the climate system.

mandatory national carbon dioxide emissions reduction target for 2050 to be 90 per cent below 1990 emissions rather than 60 per cent below, as initially proposed. The adverts appealed to the authority of science by claiming, 'Scientists have agreed that the Earth must not exceed an average temperature rise of 2°C, otherwise catastrophic climate change will be inevitable.'[3] Implicit in this appeal to science as the basis for a lobbying campaign is that science has the authority to make definitive and universal statements about what is and what is not dangerous for people and societies and, ultimately, for the world. Science is being used to justify claims not merely about how the world is (what are called 'positive' statements), but about what is or is not desirable – about how the world *should* be ('normative' statements).

In this chapter we explore whether **one of the reasons we disagree about climate change is because science is not doing the job we expect or want it to do.** Maybe we disagree about climate change because we have different expectations about what science can or should tell us, or because we view the authority of scientific knowledge in different ways. The three questions examined above – What is causing climate change? By how much is warming likely to accelerate? What level of warming is dangerous? – represent just three of a number of contested or uncertain areas of knowledge about climate change. If science is not able to adequately distinguish 'truth' from 'error', or if science is not able to offer reliable estimates of the likelihood of future climatic phenomena occurring, is the problem with science itself or with our expectations of what science can do for us? And what authority does science have when it comes to making normative judgements about individual and social risks and dangers? Is science, in this case the science of climate change, a straightforward activity concerned with discovering impartial truth about the world and then disclosing this truth to politicians so that

[3] Quoted in the *Guardian Magazine*, 29 September 2007.

truth-based policies will follow? Or are there different ways of understanding the abilities and roles of science in society? In this chapter we therefore consider the performance of science.

Science thrives on disagreement. Science can only function through questioning and challenge. It needs the oxygen of scepticism and dispute in order to flourish. Indeed, 'organised scepticism' was one of the four norms – the moral principles governing scientific enquiry – which American sociologist Robert Merton advocated in his 1942 treatise on the structure of science. These norms – the other three being 'universalism', 'commun[al]ism' and 'disinterestedness' [4] – have dominated the conduct and self-image of scientists for much of the last half-century. But disagreements about knowledge claims may often be rooted in more fundamental differences between the protagonists. Presented as disputes about scientific theory, evidence or prediction, disagreements may rather be about *how* knowledge can be established (what is called 'epistemology'), about the role of values in such knowledge-seeking, or about the role of scientific knowledge in policy making. We may disagree about climate change simply because we understand and use science in different ways.

In answering the questions posed above we will have to understand something about the nature of scientific knowledge and about how this knowledge gets used in society. We start by exploring what we mean when we talk about scientific knowledge (Section 3.2). We contrast positivist and constructivist conceptions of knowledge,[5] and

---

[4] Merton, R. K. (1942) The normative structure of science, reproduced in Merton, R. K. (1973) *The sociology of science: theoretical and empirical investigations*. University of Chicago Press.

[5] Put simply, a positivist view of knowledge holds that reality is independent of human consciousness and is external, material and objective. Because reality is 'out there', it can be studied independently of the characteristics of the enquirer. A constructivist view of knowledge holds that reality is essentially subjective, and that 'truth' is therefore a construction which is located within our own experience – historically, culturally, experientially. At its extreme, there are as many constructed realities as there are people.

also consider the relationship between 'universal' scientific knowledge about climate change and what has become known as 'local environmental knowledge'. This leads us to consider how scientific knowledge is established and, once established, how robust or stable it becomes (Section 3.3). This involves us in discussions about the role of experts' beliefs and the function of consensus in establishing the validity of scientific statements; issues that lie at the heart of debates about climate change.

We also need to examine two other dimensions of the scientific enterprise: how science is governed (Section 3.4) and how scientific knowledge is used in society (Section 3.5). Science is a human endeavour, a social process, and the practices and uses of science are therefore always conditional upon the society and culture in which these activities are situated. In these sections we consider the institution of the IPCC – one of the foremost international bodies for assessing scientific knowledge about climate change – and also three different ways in which scientific or expert knowledge gets used in decision making.

3.2 What is Scientific Knowledge?

Ideas about climate, and about the stability of climate (Section 2.5), were not the only things that were changing in the decades following the Second World War. At the very time that science was extending its reach geographically (through the creation of new international scientific collaborations and organisations) and educationally (through the massive expansion in the number of practising scientists), the ways in which the nature, practice and purpose of science were understood were also changing. In the 1950s, science could still be seen as an activity that would lead assuredly to greater understanding about how (most things) functioned – the stars, the world's oceans, the human brain. Such confidence in the project to discover objective truth offered the prize of more certain predictions about the natural world and (almost) unlimited prospects for new technologies by which

Nature could be manipulated and controlled for the improvement of human welfare.

Fifty years later, encountering such uncomplicated and unconditional optimism about science is rare. While science has continued its remarkable trajectory of growth and while, measured against many performance criteria, it has continued to be remarkably successful, there is now less public confidence in the practices and purposes of scientists. Scientists themselves – or at least those who think about science rather than those who just do it and those who govern science and fund it – are also more reflective about the enterprise. Science is (has to be) more responsive to the moods and opinions of the public. In 1962 the respected British philosopher Michael Polanyi could write boldly: 'I appreciate the generous sentiments which actuate the aspiration of [society] guiding the progress of science into socially beneficent channels, but I hold [this aspiration] to be impossible and indeed nonsensical.'[6] Less than fifty years later the context of scientific research has changed so radically that two science policy analysts could assert, without apology, that 'most scientific research, whether funded by public or private moneys, is intended to support, advance or achieve a goal that is extrinsic to science itself … to achieve beneficial societal outcomes'.[7]

It is no longer possible to see science in the autonomous, self-governed way in which Polanyi idealised it in his 1962 article 'The republic of science'. Neither is it possible to see scientific knowledge unproblematically as the neutral outcome of a steadily advancing pursuit of an objective and universal truth. The American philosopher of science Thomas Kuhn offered the most significant challenge to this traditional view of science. His 1960s work on the structure of

[6] p. 61 in Polanyi, M. (1962) The republic of science: its political and economic theory. *Minerva* 1, 54–74.

[7] p. 1 in Sarewitz, D. and Pielke, R. J. Jr (2006) The neglected heart of science policy: reconciling supply of and demand for science. *Environmental Science and Policy* 10, 5–16.

scientific revolutions did much more than (famously) introduce the language of 'paradigm shifts' into the history of science. It opened the way for later sociologists and philosophers to understand the human dimensions of scientific practice and the different social and cultural contexts in which science is undertaken. Where science is practised, by whom and in what era, affects the knowledge that science produces. Science not only has a methodology, but it also has a history, a geography and a sociology.

This changing appreciation of science and its role in society was particularly evident in the environmental sciences. As we shall see, grasping the significance of this change is essential if we are to understand what science is saying about climate change. Science is being called upon to do more than answer the questions, 'Is the world getting hotter?' and, if so, 'Why?' These are hard enough questions, which tax the observing systems, routine measurements, physical theories and computer models of conventional science. Climate change has emerged as a phenomenon which asks more demanding questions than just these. It is framed as an environmental risk of global proportions in which humans are implicated both as perpetrators and as victims and where policy decisions need to be made. Climate change has become a classic example of what philosophers of science Silvio Funtowicz and Jerry Ravetz have termed 'post-normal science': the application of science to public issues where 'facts are uncertain, values in dispute, stakes high and decisions urgent'[8] (see Figure 3.1). Normal science, science as guided by Merton's four classical norms of scientific practice – scepticism, universalism, communalism, disinterestedness – is no longer fit for such purpose.

In cases such as climate change, then, science must take on a different form. As well as seeking to observe, theorise and model in the quest to establish 'facts' – to formulate what is known – science must

---

[8] Funtowicz, S.O. and Ravetz, J.R. (1993) Science for a post-normal age. *Futures* 25, 739–55.

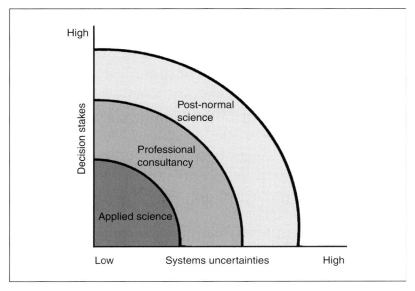

FIGURE 3.1: *Schematic representation of how post-normal science can be located in relation to more traditional problem-solving strategies. As 'system uncertainties' and 'decision stakes' increase, one moves from the realm of 'normal' applied science, where expertise is fully effective, through professional consultancy, where skill, judgement and sometimes even courage are required, to the case where risks cannot be quantified or when possible damage is irreversible. This latter realm requires post-normal science.*
Source: *Funtowicz and Ravetz (1993).*

pay much more attention to establishing and communicating what is unknown, or at best what is uncertain. In addition to striving to eliminate bias and prejudice from the practice of science – to adhere to Merton's norm of 'disinterestedness' – scientists must also recognise and reflect upon their own values and upon the collective values of their colleagues. These values and world views continually seep into their activities as scientists and inflect the knowledge that is formed. The goal of 'disinterestedness' can still be aspired to, whilst simultaneously and crucially recognising that, in failing to achieve it, scientific knowledge will inevitably take on a different character.

Funtowicz and Ravetz offer two more characteristics of 'post-normal' science: high stakes and urgent decisions. Because the stakes about climate change *are* high, science must recognise that the processes whereby knowledge is created must be open to public scrutiny and engagement. And since policy decisions about climate change are being made now, and being made continuously, the ways in which knowledge is condensed, packaged and brought into social discourse also need to be re-thought. The extent to which climate change science has been successful in these adaptations – acknowledging uncertainty, exposing values, earning trust, and speaking in society – will be examined later in the chapter.

One further concession that Polanyi's 'republic of science' has to make in relation to climate change is in respect of expertise. The 'post-normal' character of climate change demands that a wider range of expert voices must be heard in public arguments. The expertise of scientists and the claims of scientific knowledge do not exhaust the sources of expertise or authority to which society may turn in seeking guidance for the decisions that must be made. We will see this in later chapters in relation to values, beliefs and political ideologies, but here – in relation to the role of science – there are two insights that must be offered.

The definition of who counts as an expert in the creating, handling and communicating of scientific knowledge needs careful scrutiny. Or, as philosopher of science Harry Collins puts it, how do we distinguish between 'those who know what they are talking about' and those who don't?[9] The scientific academy carefully polices its membership and its internal debates through the conventions of formal qualifications, election to professional societies, and through the operation of the peer-review system. But who determines the research questions to be addressed by science, who evaluates the utility of research completed, and who selects which experts are to speak in the public sphere?

[9] p. 2 in Collins, H. and Evans, R. (2007) *Rethinking expertise.* University of Chicago Press.

For example, when Al Gore presents scientific claims about climate change in his film *An Inconvenient Truth* is he speaking as a politician, a lay expert or as a spokesperson for science – and does it matter if, as has been claimed,[10] he gets some aspects of the science wrong? In recent years, these questions of expertise with respect to climate change have become more contentious if only because of the 'high stakes' and 'urgent decisions' characteristics of post-normal science. In order to democratise science and to give greater credibility to the contract that exists between science and society, some have advocated for greater lay expertise to operate in the governance of science. The supposed benefits of including lay expertise in the governance of science include 'ensuring researchers are addressing the right questions, incorporating the knowledge of these non-academic experts in the analysis, and testing the validity and practicality of any prescriptions researchers are proposing'.[11] This is a long way from Polanyi's (1962) self-governing republic.

Not only must science concede some of its governance to wider society, it must also concede some ground to other ways of knowing. As we saw in Chapter 1: *The Social Meanings of Climate*, there are ways of understanding climate other than as a globalised, physical entity observable through scientific measurement and manipulated in computer models. As climate change discourses have reached out geographically and culturally, the value and role of local environmental knowledge about climate has begun to be recognised. Such local knowledge – sometimes referred to as indigenous or traditional knowledge – has frequently been established over centuries of habitation and is often unique to a particular community or ethnic group.

---

[10] In October 2007, in a case brought before Britain's High Court of Justice, a ruling was made that the film *An Inconvenient Truth* contained nine 'scientific errors' and should therefore not be distributed to UK schools without accompanying 'corrective' guidance material for teachers.

[11] p. 829 in Scott, A. (2007) Peer review and the relevance of science. *Futures* 39, 827–45.

In relation to climate, such local tacit knowledge might entail, for example, indigenous strategies for drought management, local monitoring of environmental change, or traditional modes of weather forecasting. Although not conventionally classified as scientific knowledge, recent years have seen a number of efforts to bring together scientific and indigenous ways of understanding the natural world and climate change. This has been especially evident in northern Arctic communities and in some African dryland communities. Local environmental knowledge presents a challenge to another of Merton's norms – that of 'universalism' – and questions the automatic instinct in many Western societies to associate knowledge expertise with the community of scientists.

Science is a human enterprise and therefore has its own rules, procedures and norms, its own particular ways of knowing. These rules can change over time and, as we have seen, have indeed done so. As sociologists Bruno Latour and Steve Woolgar showed in their groundbreaking 1979 book *Laboratory Life: The social construction of scientific facts*, science and the community of scientists can be studied in much the same way as anthropologists study the norms and practices of different ethnic communities. The relationship between science and society has also changed in recent years. There is an increasing appreciation, both among scientists and among the public, of the contingent factors of personal belief, cultural context and institutional arrangements, which influence the way scientific knowledge is established. Somewhere in between science as the sublime discovery of absolute truth – a purely positivist reading of science – and science as a hopelessly subjective activity – a purely constructivist mentality – is a more nuanced understanding of what science can do and what it can't do and a deeper understanding of the nature of scientific knowledge. Essential to this new understanding is an appreciation of scientific uncertainty. Far from being able to eliminate uncertainty, science – especially climate change science – is most useful to society when it finds good ways of recognising, managing and communicating uncertainty.

3.3 How Robust is Scientific Knowledge?

Uncertainty pervades scientific predictions about the future performance of global and regional climates. And uncertainties multiply when considering all the consequences that might follow from such changes in climate. Some of this uncertainty originates from an incomplete understanding of how the physical climate system works – the effects of atmospheric aerosols on clouds, for example, or the role of the deep ocean in altering surface heat exchanges. In principle, many of these uncertainties may be reduced over time, or at least quantified formally. Other sources of uncertainty emerge from the innate unpredictability of large, complex and chaotic systems such as the global atmosphere and ocean. At best, we may be able to estimate some probability of a particular future climate outcome occurring – as is done routinely with weather forecasts. A third category of uncertainty originates as a consequence of humans being part of the future being predicted. Individual and collective human choices five, twenty or fifty years into the future are not predictable in any scientific sense. Here, the best that can be done is to work with a range of broad-scale scenarios, a range of possible futures.

Collectively, these uncertainties present serious problems for politicians, policy makers and strategic decision makers needing to weigh the risks of future climate change in decisions to be made today. Although scientific predictions about arcane or merely philosophical questions can afford to be nonchalant about uncertainty, the 'high stakes, urgent decisions' character of climate change means that policy makers do not have the luxury of nonchalance. They are faced with the problem of making sensible decisions today in the face of uncertain risks tomorrow. There is a demand for knowledge about future climate risks and a demand to know the costs and benefits of different courses of action with respect to the economy, to natural resources and to social well-being.

Yet the curious thing about the current relationship between climate change science, knowledge and decision making is that it should be 'uncertainty' rather than 'certainty' that is the condition perceived to be anomalous. The assumption seems to be that certainty is an attainable state of knowledge, despite the evidence that for most of human history we have accepted that *lack* of certainty is our natural lot. As sociologist of science Sheila Jasanoff explains: 'Uncertainty has become a threat to collective action, the disease that knowledge must cure. It is the condition that poses cruel dilemmas for decision makers; that must be reduced at all costs; that is tamed with scenarios and assessments; and that feeds the frenzy for new knowledge, much of it scientific.'[12]

Science itself has been partly responsible for this change in expectation with respect to climate change – the new notion that certainty is attainable – since science has opened up the possibility for humans, for the first time, to know something about the long-term climate future. Given this state of affairs, it is incumbent on scientists to indeed try to 'tame' the uncertainties which plague their predictions of the climatic future. We will explore here two ways in which scientists seek to do this: first, by recognising the importance of subjectivity in scientific knowledge creation, for example by adopting what is formally referred to as a Bayesian belief system; and, second, by accepting consensus-building as a means to establish scientific knowledge. In both cases, the attempt to reveal and manage uncertainty challenges Merton's norms of 'universalism', 'disinterestedness' and 'organised scepticism'; norms which for a long time past, and still commonly today, are viewed as the distinguishing hallmarks of a scientist.

Bayesian beliefs

When knowledge about the processes determining a future outcome is uncertain, and hence where the likelihood of that outcome occurring is uncertain, it is legitimate for scientists to express their subjective level

---

[12] Jasanoff, S. (2007) Technologies of humility. *Nature*, 450, 33.

of confidence about a given outcome. In other words, it is legitimate for them to reveal their degree of belief that a prediction will turn out to be correct. If it is not certain whether an event will occur or not, something has to substitute for this lack of certainty – because the only alternative is to admit that absolutely nothing is known and say nothing at all. Statements of subjective probabilities or beliefs are known as Bayesian, after the eighteenth-century Scottish mathematician and Presbyterian minister Thomas Bayes, who developed the first modern understanding of probability. Bayesian probabilities of an event occurring are in effect an individual's (or a group of individuals') estimate – their belief or subjective judgement drawing upon evidence and experience – of the likelihood of that event occurring.

In one sense, Bayesian statements are no more or less than informed judgements about likelihoods, much as any one of us might make regarding the chance of our own nation's football team winning the next World Cup. The reliability or robustness of these (knowledge) statements about outcomes depends on two main factors. The first of these is how well constrained is the outcome by known cause-and-effect processes, knowledge about which we have good reason in which to place our confidence. The second is how well qualified is the person making the Bayesian statement to understand how those known processes are operating in any particular case. Let us illustrate these two dimensions of subjective knowledge.

The likelihood of the sun rising tomorrow is well constrained by a number of physical processes, in the understanding of which I place great confidence. In contrast, there are too many factors about which I know very little that will influence whether or not England wins the next World Cup. My Bayesian belief about the sun rising – virtually certain – reveals a much higher level of confidence that does my belief about England winning the World Cup – deeply unsure. Scientific understanding of physical cause and effect therefore really does matter.

Let us consider the second dimension of Bayesian beliefs using a different example. Although the success of a particular course of medication prescribed for me may be quite well constrained by biophysical processes about which members of the medical profession have some considerable knowledge, I would place much greater confidence in the belief expressed by my doctor that a particular medication will cure me than I would in my own belief about the effectiveness of the treatment. Relevant expertise therefore also really does matter.

For many aspects of the climate future, and of the consequences of those climate futures, scientists end up having to express their knowledge using a Bayesian framework. It matters that the events they are describing – the probability of heatwave frequencies doubling or hurricane intensities increasing – are well constrained by physical processes which can be understood using the basic laws of physics. It also matters that the people expressing these judgements are experts about heatwaves or about hurricanes. Thus the IPCC Fourth Assessment Report concluded that there was a more than 90 per cent chance that heatwaves would become more frequent during this century, and a more than 66 per cent chance that hurricanes would become more intense. These Bayesian statements are statements about the degree of belief that the experts hold that the events will occur, given the observations, physical theory and modelling results that are currently available to them.

Science has established a number of formal methods and rules for ensuring that this 'organised subjective' approach to knowledge and uncertainty is well managed. One such method is that of expert elicitation. Rather than revealing the beliefs of one expert about a particular outcome, an expert elicitation will allow several experts – usually ten or more – to systematically quantify their beliefs about the likelihood of this outcome occurring. This expert elicitation approach was used for the first time in relation to climate change in 1978, when the US National Defense University (NDU) in Washington DC published a report, *Climate Change to the Year 2000*. Nineteen panellists expressed their beliefs about whether world temperature would increase, decrease or stay the same by the year 2000. Strict protocols should operate in such exercises and the resulting expressed beliefs may be displayed as a range of values and/or reduced to a central tendency. In the NDU study cited above, rather than seeking a consensus judgement through dialogue, the expert views were statistically aggregated to offer a single consolidated judgement that by the end of the century 'a climate resembling the average for the past thirty years'[13] was most likely. The NDU study was later criticised for not paying enough attention to the reasons for disagreement among experts and, as climate unfolded over the following two decades, the judgement turned out to be inaccurate.

Which brings us to the second way in which uncertain knowledge can be managed and communicated – through consensus.

Consensus in science

At first sight using consensus – 'an opinion or viewpoint reached by a group as a whole, usually by majority' – as a means for establishing scientific knowledge seems at odds with what science claims it is good at doing. It seems a long way from the idea of science as an objective adjudicator between 'truth' and 'error'. But, as we have seen with Bayesian beliefs, the use of consensus is merely one (structured) way of distilling evidence – evidence which might be somewhat ambiguous, incomplete or contradictory or where there is latitude for genuine differences of interpretation – into an overall agreed statement on an issue of scientific or public importance.

Consensus in science is normally achieved through communication at conferences and through the processes of peer-review and publication. Scientific institutions or professional associations may issue consensus position statements in order to communicate a summary of agreed scientific knowledge to a wider public audience. For

[13] p. iii in NDU (1978) *Climate change to the year 2000*. National Defense University: Washington DC. The methods employed in this expert elicitation were later criticised by Schneider, S. H. (1985) Science by consensus: the case of the NDU study 'climate change to the year 2000' – an editorial. *Climatic Change* 7, 153–7.

example, the American Meteorological Society formally adopted an agreed *Information Statement on Climate Change* in February 2007, intended to provide a 'trustworthy, objective, and scientifically up-to-date explanation of scientific issues of concern to the public at large'. In cases where there is little controversy regarding the subject under study, establishing a scientific consensus can be quite straightforward. In other cases – such as climate change – arriving at a consensus position can be much more complicated. The largest international scientific assessment of climate change – the IPCC – has frequently adopted a consensus mode of establishing and communicating knowledge. These consensus statements have not always been without controversy, attracting accusations both for being too radical on aspects of the science and for being too cautious in estimating risks (see Box 3.1 for an example).

Box 3.1: The Limits of Consensus: The IPCC and the Greenland Ice Sheet

The response of the Greenland Ice Sheet (GIS) to changes in climate is important for two main reasons. The ice sheet contains enough water to raise the average level of the world's oceans by about six metres if it fully melted or discharged into the northern oceans. Furthermore, the injection of such large volumes of fresh water into these saline waters might contribute to a weakening of the thermohaline circulation in the world's oceans; a circulation which at present conveys the warm Gulf Stream into the waters surrounding northwest Europe, thereby moderating the regional climate.

In the Fourth Assessment Report (AR4) of the IPCC, the expert scientists considering the behaviour of the GIS all agreed that climate warming would lead to ice loss to the oceans over the coming century. The precise mechanisms by which this ice loss could *substantially* accelerate – i.e. through surface melting and through enhanced glacier flow due to removal of retaining ice shelves or to sub-glacial meltwater – were not well enough understood, however, to be adequately simulated in computer models of the ice sheet's dynamics. The overall consensus of the AR4 was that there was less than a one-in-twenty chance that the global sea level would rise this century by more than 59 cm and that the GIS would contribute up to a maximum of about 10 cm to this total. Rather than put any numbers on the consequences of an *acceleration* of Greenland ice melt, the IPCC experts said that 'understanding of these effects is too limited to assess their likelihood or to give a best estimate'.[14]

After the report was published, this conclusion caused a round of commentaries to be written by a number of climate scientists, some of whom had themselves been part of the IPCC expert writing team, in which concern was expressed at the way in which this consensus position had been reached. The range of the projected IPCC global sea-level rise by 2100 had narrowed from between 9 and 88 cm in the Third Assessment Report, published six years earlier in 2001, to between 18 and 59 cm in the Fourth Assessment Report. The IPCC stated that this narrowing of the range was due to improved understanding of some of the uncertainties affecting future sea level. However, as the critics of the IPCC pointed out, the most important uncertainty about the GIS and its contribution to sea-level rise – accelerated ice flow due to sub-glacial dynamical processes – had *not* been quantified and had therefore been excluded from the calculated estimates. Their argument was that the IPCC consensus range of 18–59 cm sea-level rise by 2100 was not conveying to policy makers, or to the public, the full range of uncertainties which scientific research had revealed.

---

[14] IPCC, *Climate change 2007*, p. 71.

This example well illustrates the difficulties that scientists face when producing consensus statements for public consumption based on expert knowledge. Such statements are required to be well founded on evidence, to carry the authority of scientific experts, and to balance the tension between what is known, believed likely and quantifiable against what is unknown, only possible and qualitative. In this case, the AR4 tended to place greatest weight on the quantitative results from models of the Greenland Ice Sheet, implicitly arguing that such models best capture existing knowledge that is both robust and quantitatively tested. As one expert claimed: 'We just don't have the capacity to quantify [that sort of ice-sheet behaviour] ... the best you can do is point some red flags ... the language of the IPCC does just that.' [15]

On the other hand, if these physically based models exclude complex ice-sheet processes which are believed to be relevant, then their quantitative results may underestimate potential future increases in ice melt rates. Certainly, in the eyes of some scientific critics, the consensus process of the IPCC is lacking something: 'There are limitations on that process. Everybody in [sea-level] research is much more concerned than six or seven years ago ... there is a role for something in addition.' [16] One option would be to establish a more formal expert elicitation in which the Bayesian beliefs of experts would be revealed and collated. Such a route to establishing the likelihood of future climate risks would place greater weight on the formally expressed qualitative judgements of experts than on the quantified results of numerical models. How risks are determined, perceived and communicated – by expert and non-expert alike – is something that we shall look at in Chapter 6: *The Things We Fear.*

---

[15] p. 1413 in Kerr, R. (2007) Pushing the scary side of global warming. *Nature* 316, 1412–15.

[16] *Ibid.*, p. 1415.

The example of the Greenland Ice Sheet illustrates how experts can disagree about climate change, certainly how they can disagree about the risks posed by climate change. This may be because of differences in the way evidence is interpreted and in the way confidence in the available evidence – models, theory or observation – affects the judgements of individual scientists. What is clear – whether in the adoption of Bayesian or consensus modes of knowledge production – is that scientific knowledge is conditional, contains inescapable elements of subjectivity, and can change as beliefs (for example about the credibility of evidence) change. In this respect, for policy makers to accept the scientific consensus of the day can prove dangerous in some situations; for instance when making substantive policy decisions based on theories or risk assessments which later turn out to be false. For example, the compulsory sterilisation of thousands of mentally ill patients in the USA during the 1920s and 1930s was mandated under the false (consensus) notion that it would end mental illness.[17] Additionally, because of the flexible relationship between knowledge and judgement implicit in such methods, it is easy for critics to emphasise the subjective nature of knowledge in general and to argue that science is being used to justify *a priori* opinions.

On the other hand, where stakes are high and decisions urgent, some way has to be found of bringing scientific evidence into the policy-making process even when uncertainties remain endemic and even when there is not unanimity among all scientific experts. In such circumstances, expert elicitation and/or large consensus-driven scientific assessments are probably the 'least worst' way of progressing. A consensus approach allows for a range of different beliefs and judgements to be recognised. Indeed, reaching a consensus implies, by definition, that a range of different views exist and are expressed. Critics of the IPCC, whichever angle they come from, are correct in

---

17 Kevles, D. (1985) *In the name of eugenics: genetics and the uses of human heredity.* Knopf: New York.

saying that mere consensus does not establish 'the facts' or affirm 'the truth'. Nor does consensus guarantee that the assessment of future risks is stable; the consensus about what is deemed to be true may change over time, even if 'truth', in some absolute sense, does not. Yet if the process of consensus building is open, transparent and well governed, if it seeks to be true to the many uncertainties which persist, then consensus may offer policy makers the best that science has to offer. And as we shall see in Section 3.5, establishing the credentials of any scientific knowledge claim carries no inevitable implications for the subsequent negotiation and adoption of any particular policy.

What consequently becomes essential for consensus knowledge to carry authority and relevance into the public sphere and into decision makers' deliberations are the processes of governance, through which consensus is consolidated and by which disputes – Merton's 'organised scepticism' – are handled. It is to these issues of governance that we now turn.

3.4 How is Science Governed?

We have already seen how the changing nature of science, and its changing relationship with society, has left Polanyi's self-governing republic of science somewhat defenceless. The boundaries of science have become more porous even if, in the heartland of the republic, the norms and practices of expert peer-review and disinterested enquiry continue to be aspired to. New complex phenomena such as climate change, and the role ascribed to scientific knowledge in public debates about climate change policy, demand adjustments to the way in which science is governed and how its knowledge is policed.

With respect to climate change, the Intergovernmental Panel on Climate Change is the creation of this new operating environment for science or, one might say, is an attempt to create a new interface between science and policy suited to a 'post-normal' operation of science. Yet exactly what the IPCC is, how it is governed, what sort of knowledge it produces, and with what authority its knowledge is endowed are all matters of some ambiguity (see Box 3.2 for an example of this).

Box 3.2: The Royal Society and ExxonMobil

On 4 September 2006, the Royal Society, the UK's national science academy and the oldest scientific academy in the world, wrote a letter to ExxonMobil, the world's richest corporation, challenging its views on climate change and its funding of some scientific institutions. The exchange of letters was made public and a flurry of media reporting ensued in the UK and beyond. The exchange of views well illustrates the way in which different readings of scientific knowledge lead to disagreements about climate change.

The Royal Society's letter challenged Exxon with regard to its interpretation of the status and robustness of the conclusion of the IPCC that 'growing confidence' can be placed on the attribution of recent warming to greenhouse gases. In its Annual Report, Exxon had claimed that this IPCC conclusion relied on 'expert judgement rather than objective, reproducible statistical methods', and that consequently it was very difficult to determine objectively the extent to which recent climate change might be the result of human actions.

The objection of the Royal Society centred on the interpretation of the authority of the IPCC's scientific assessments. In its letter to Exxon, the Society claimed that Exxon's statements were misleading and that 'the "expert judgement" [quote marks in the original] was actually based on objective and quantitative analyses and methods ... which have been independently reproduced'.[18] In their

---

[18] Letter from Bob Ward, Senior Manager, Policy Communication, at the Royal Society, to Nick Thomas, Director of Corporate Affairs, Exxon UK, on 4 September 2006.

reply a few weeks later, Exxon claimed that their views had been misrepresented by the Society and that their Annual Report stated that, 'even with many uncertainties, the risk that greenhouse gas emissions may have serious impacts justifies taking action'.[19] Exxon also defended themselves against the charge that of the sixty-four organisations they had funded for 'public information and policy research', thirty-nine were featuring information on their websites that, according to the Royal Society, 'misrepresented the science of climate change'.

This dispute about the status of knowledge about climate change hinged around the differing attributions of authority to the role of expert judgement in the IPCC reports. Exxon were claiming that expert judgement was inferior to the use of 'objective methods' and, implicitly, was an inadequate basis for knowledge. The Royal Society, however, claimed that the expert judgement was in fact based on reproducible 'objective methods' and therefore there was no distinction to be drawn between the two approaches (presumably this was why the Society placed the term 'expert judgement' in inverted commas).

Two things in particular are interesting about this exchange. The first is that these two large and powerful organisations chose to fight a battle of viewpoints – a battle which ended up being conducted in public – over interpretations of the meaning and legitimacy of 'expert judgement' in the formation of scientific knowledge. Exxon implied that expert judgement is inferior to 'objective methods' as a route for obtaining and presenting knowledge, which the Society indirectly agreed with by basing their ultimate defence of expert judgement also on 'objective methods'. Both organisations therefore rested their ultimate knowledge claims on 'objective methods'

---

[19] Letter from ExxonMobil Vice-President, Kenneth Cohen, to the President of the Royal Society, Lord Rees, on 25 September 2006.

but, as we have seen in the main text, in complex and uncertain areas of knowledge 'objective methods' alone are rarely adequate to establish what is known. The two organisations seem to differ only in the way in which the results of these 'objective methods' are distilled, placed in narrative form and brought to a wider audience. If this isn't to be done by expert judgement, one wonders how else Exxon would prefer this to be accomplished?

The second point of interest here is the role seemingly taken on by the Royal Society as an enforcement authority acting on behalf of the IPCC and of the scientists whose work is represented in the IPCC. The Society asked Exxon to indicate whether they would be continuing to express views that were 'inconsistent with the findings of the work' of Fellows of the Society, although with what ultimate purpose this request was made was not clear. The question raised here is who speaks for science and whose role is it to correct 'misrepresentation' – the individual scientist, the Society as a professional body of elite scientists (a sort of trades union), or the IPCC who were ultimately responsible for mandating the expert judgement in the first place? Policing the representation of scientific knowledge is a contentious activity.

The IPCC was established in 1988, initially under the auspices of the World Meteorological Organization which was quickly joined by the United Nations Environment Programme. It was mandated by the UN General Assembly to undertake international assessments of scientific knowledge about climate change, its impacts, and the range of possible response strategies. It was to be governed by a Bureau consisting of selected governmental representatives, thus ensuring that the Panel's work was clearly seen to be serving the needs of government and policy. The Panel was not to be a self-governing body of independent scientists.

From its very beginning the IPCC has acted as a hybrid or 'boundary' organisation, operating as it does across the worlds of science, public discourse and policy. The IPCC was charged with producing policy-relevant assessments of knowledge, without these assessments claiming to be policy-prescriptive. As stated in its mandate, 'the IPCC does not recommend policies'. Its function was to be advisory and not formative, seeking to keep some clear distinction between the worlds of science and policy. Thus, 'all of these [designated] activities went a long way toward bringing the Panel's institutional apparatus into line with the increasingly accepted view of the organisation as a technical advisory body for the formation of global policy'.[20]

But this distinct boundary between science and policy has proved a difficult one to maintain and to police. Because the boundary is often fluid, different actors and advocates in public debates about climate change have felt entitled to make different and contradictory claims about the nature and authority of IPCC's pronouncements. Does the IPCC offer impartial science or does it shape policy? Does the IPCC reflect the views of the participating scientists or the views of the government officials who have to approve its reports? Some disagreements about climate change can be traced to different interpretations of the authority of the IPCC. We will mention three of the more visible ones: the validity of consensus, the legitimacy of experts and the neutrality of advice.

We have discussed earlier (see the section on Consensus in science) why it may often be necessary for science to use consensus processes as a way of consolidating knowledge so that it can be useful for policy. Consensus knowledge, by construction, will always allow experts to disagree, with knowledgeable opinion existing at either tail of the distribution of views (Box 3.1 gives an illustration of this). Such scientific

---

[20] p. 61 in Miller, C. A. (2004) Climate science and the making of a global political order, in Jasanoff, S. (ed.), *States of knowledge: the co-production of science and the social order*. Routledge: London, pp. 46–66.

consensus is not ultimate 'truth' and, on occasion, may turn out to be wrong. But the alternatives to the IPCC style of consensus-building are even less likely to command widespread authority within the worlds of science and policy. 'Vastly better [than random solicitation of views] is the work of groups such as the IPCC … which although slow, deliberative, sometimes elitist and occasionally dominated by strong personalities, are nonetheless the best representation of the scientific community's current general opinion.'[21] It is also claimed that consensus-building will always tend to produce conservative outcomes, hedging against the twin dangers of claiming there is a risk when there isn't one and of ignoring a risk when there is one. These limitations to consensus knowledge must be openly acknowledged and must be represented in communications about the relationship between science and climate change. How this is done is itself a source of contention that we will look at in Chapter 7: *The Communication of Risk*.

The second example of contention concerns the expertise of those involved in the IPCC. Forming consensus knowledge places great weight on the selection of the experts who are asked to negotiate the consensus and, in the case of the IPCC, the experts who are invited to review and critique this knowledge during its formative stages. Here too, the IPCC must negotiate its way through a difficult set of procedures and judgements. Criticisms are made that the experts are not independent, since they are nominated by governments; that they are biased towards the scientifically elite nations of the North and West; that some forms of expert knowledge – local environmental knowledge, or certain academic disciplines such as sociology or anthropology – are excluded; or that the peer-reviewers of the IPCC

---

[21] pp. 244–5 in Edwards, P. N. and Schneider, S. H. (2001) Self-governance and peer review in science-for-policy: the case of the IPCC Second Assessment Report, in Miller, C. A. and Edwards, P. N. (eds), *Changing the atmosphere: expert knowledge and environmental governance*. MIT Press: Cambridge, MA, pp. 219–46.

draft texts represent government, business or environmental advocacy interests and are therefore not experts.

Each of these criticisms carries some weight. There is no self-evident group of 'international climate change experts' and who is or is not part of the IPCC is often poorly defined and constantly changing.[22] Social scientists in general are indeed poorly represented among the nominated experts, and elite judgements clearly are made about inclusivity and participation. But the rules and procedures of engagement, and the transparency of these rules and procedures, perhaps suggests that the IPCC is – as in Winston Churchill's famous aphorism about democracy as a form of governance – the worst of all possible ways of assessing knowledge about climate change … apart from all the others. Nevertheless, the fluid membership of the IPCC's writing and reviewing teams does make it rather difficult to know who exactly it was who won the 2007 Nobel Peace Prize.[23]

A third area of contention concerns the distinction between scientific knowledge which is policy-neutral and knowledge which is policy-prescriptive. Following Merton's classic norm of scientific 'disinterestedness', it may be easy to suggest that such a clean distinction can be maintained. Yet for intergovernmental assessments such as the IPCC always to recognise this distinction, even if it could be policed, is not easy. US policy analyst Roger Pielke has identified four roles that scientists adopt in relation to policy making: pure scientist

---

[22] On several occasions experts have resigned from IPCC author teams during the drafting stages because they disagree with the consensus knowledge that is forming; other experts may be added during the process and, as with all collaborative efforts, some experts are very vocal and others are 'sleepers'.

[23] The Norwegian Nobel Prize Committee jointly awarded the 2007 Peace Prize to the IPCC, stating: 'Through the scientific reports it has issued over the past two decades, the IPCC has created an ever-broader informed consensus about the connection between human activities and global warming. Thousands of scientists and officials from over 100 countries have collaborated to achieve greater certainty as to the scale of the warming.' These scientists and officials were never named by the Nobel Committee.

(Merton's disinterested scientist), science arbiter, issue advocate, and honest broker.[24] To act always in the role of honest broker – 'expanding the choice of options for policy makers' – rather than as an issue advocate – 'using science to promote a specific policy' – may be hard to do. It may be even harder always to be *perceived* as so doing. Some of the criticism of the IPCC – some of the reasons that we disagree about climate change – are related, for example, to the perception that the IPCC endorses the European Union policy goal of keeping global warming to no more than 2°C above the pre-industrial level or that it endorses the Kyoto Protocol as the only framework through which international policy should be formed. Although a careful reading of the IPCC reports would not support such perceptions, there are strong temptations for such advocacy positions to be adopted by individual contributors to, or commentators on, the IPCC reports. In the mind of the public, such nuanced distinctions may be hard to appreciate.

3.5 How is Scientific Knowledge Used in Society?

The examples in the previous section of how scientific knowledge about climate change is constructed by the IPCC and how such knowledge is regarded as it enters the wider social world, demonstrate how hard it is to navigate the dangerous waters surrounding the boundaries of the republic of science. Far easier would be for science to remain far away from such trouble in its rarefied and autonomous heartland. But those days for science – certainly for climate science – have long gone. Science is clearly called upon to speak in and contribute to public and policy debates about climate change and, as it does so, it struggles to find new institutional forms and processes to shape knowledge into a usable form. By taking on this challenge, science also finds itself

---

[24] See Pielke, R. Jr (2007) *The honest broker: making sense of science in policy and politics*. Cambridge University Press.

subject to new forces which can reshape its knowledge and alter its character. As climate science rubs up against society, the nature of scientific knowledge about climate change is modified. And as specific climate policy responses are proposed, challenged and negotiated within and between nations, how scientific knowledge is viewed by society also changes.

Some of the confusion here, part of the reason that we disagree about climate change, is that there are a number of different models of how science is (or should be) used in policy development. These differences are not often made clear in public discourse; indeed the protagonists themselves might not even be aware of what model they are operating under. Differences in these operating models exist within and between communities of scientists, the cadres of national policy makers, members of the agora (public debating spaces), and the civic cultures of nation-states. Understanding and recognising these different models of science–policy interaction is crucial if we are to make progress in understanding why different advocates interpret knowledge claims from the IPCC differently, and if we are to understand some of the reasons that we disagree about climate change. I will summarise here three of the more common science–policy models in circulation and illustrate their significance with respect to the question of what constitutes 'dangerous climate change' and its policy implications.

The 'decisionist' model

One view of how knowledge relates to policy, probably the one with the longest history, is what has been christened by German sociologist Max Weber the 'decisionist' model.[25] In this model, policies can never be determined by 'facts' alone and so it is the responsibility of (elected or appointed) politicians to determine the *ends* of policy and for experts to then consider and evaluate the various *means* available to achieve

---

[25] Weber, M. (1946) *Essays in sociology* (trans./eds H. H. Gerth and C. W. Mills). Oxford University Press.

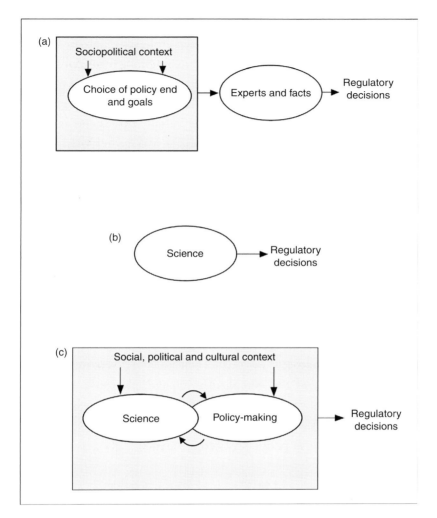

FIGURE 3.2: *Three science–policy models illustrating different relationships between expert (scientific) knowledge and regulatory decision making: (a) 'decisionist' model; (b) 'technocratic' model; (c) 'co-production' model. Source: Redrawn from Millstone (2005)*

those ends (Figure 3.2a). The deliberations and judgements of scientific or technical experts follow from the judgements of the politicians. The politicians are accountable to the electorate; the experts are accountable to the ministers who appointed them to the expert committees.

If we use this model to illustrate the question of dangerous climate change, then it would be for governments, perhaps through the United Nations, to determine what level of climate change was dangerous. Scientists and other experts would then assess the ways and efficiencies with which different policies could deliver the goal of restricting climate change to that level. To some extent, this is the science–policy model which has been followed in the European Union, which decided in 1996 to adopt the policy goal of limiting global warming to no more than 2°C above pre-industrial levels. This was largely a political decision, and since then many EU experts have been called upon to devise and evaluate different packages of policy measures that could deliver that goal.

The 'technocratic' model

The 'decisionist' model is considered to have emerged originally in industrial societies in the nineteenth century, when public policy issues were relatively simple and when scientific knowledge was still relatively uncomplicated. As science during the twentieth century increasingly revealed the complexities of multiple cause and effect that lie behind many public risks, a different model for how knowledge related to policy was adopted – the idea of a 'technocracy' (Figure 3.2b). In this approach, the politician becomes increasingly dependent on the expert for the development of policy, a view encapsulated in the claim that policy should be based on 'sound science'. US President George H. Bush reflected this view in a speech from 1990: 'Science, like any field of endeavour, relies on freedom of enquiry, and one of the hallmarks of that freedom is objectivity. Now, more than ever on issues like climate change and AIDS research … government relies on the impartial perspective of science for guidance.'[26]

---

[26] President George H.W. Bush, 23 April 1990, quoted by the Union of Concerned Scientists. www.ucsusa.org/scientific_integrity/abuses_of_science/scientists-sign-on-statement.html [accessed 9 July 2008].

This science–policy model, widely accepted today by the public, politicians and scientists, is founded on a classic view of discoverable and objective scientific 'facts', which are socially and politically neutral, and the belief that all the relevant facts can be revealed by science. It requires a skilled and compliant technocracy offering their impartial scientific knowledge to policy makers. The aphorism 'truth shall speak to power' captures its essence. In the context of our illustration, this model would suppose that the threshold level of dangerous climate change can be discovered by scientists; if not by dedicated scientific research programmes then at least implicitly through an international scientific assessment such as the IPCC. Once discovered, the responsibility then moves to the policy makers to develop and implement policy and regulations to ensure that this discovered 'scientific' threshold will not be breached. Technocratic models of knowledge-for-policy can be very attractive for politicians because the ultimate responsibility for the policy goal resides with their expert advisors, the scientists, rather than with the judgement of the politician.

It was in this spirit that, in February 2005, the UK Government hosted an international science conference on Avoiding Dangerous Climate Change which was mandated by Prime Minister Tony Blair to 'identify what level of greenhouse gases in the atmosphere is self-evidently too much'. The scientists would speak, danger would be revealed, policy would follow. This is the same model adopted by the organisation Christian Aid in their advertising campaign quoted at the beginning of this chapter: the scientists have spoken about what level of climate change must be avoided and so policy must surely follow. A technocratic model for relating science to policy places a high premium on scientists being trusted and getting the science 'right'. Campaigners operating under this model will therefore complain if they feel that scientists are selling them short, as illustrated by this public plea from British environmentalist George Monbiot and four other leading environmental campaigners in 2005: 'We're not asking you [scientists] to become campaigners or to compromise your independence. But we wish you would defend your profession [from the attack by sceptics] as any other professionals would … isn't it time you started fighting for your science?'[27]

Science becomes the battleground over which different advocacy groups will fight, because in this technocratic model winning the fight over science is equivalent to winning the fight over policy. The stakes become very high.

The 'co-production' model

Neither a 'decisionist' model – the goals set by politicians – nor a 'technocracy' model – the goals set by experts – does justice to the growing complexity of the relationships between politicians, citizens and scientists in the face of intractable and pervasive risks such as climate change. Although it is a view still widely held, the clean separation between scientific fact and policy judgement, whether operating in either a 'decisionist' or 'technocracy' framework, has become very difficult to justify. Policy making is very obviously a product of both scientific and policy judgements, under scrutiny from an increasingly active and vocal citizenry which is suspicious of leaving important decisions about public safety to either politicians or experts, or to both. Thus Sheila Jasanoff observes: 'Studies of scientific advisors leave in tatters the notion that it is possible, in practice, to restrict the advisory process to technical issues or that the subjective values of scientists are irrelevant to decision making.'[28]

In this 'co-production' model of knowledge and policy there is recognition that both the goals of policy and the means of securing those goals emerge out of joint scientific and non-scientific (i.e. political or value-driven) considerations (Figure 3.2c). The way to consider dangerous climate change under this model of science–policy

---

[27] p. 559 in Monbiot, G., Lynas, M., Marshall, G., Juniper, T. and Tindale, S. (2005) Time to speak up for climate-change science. *Nature* **434**, 559.
[28] p. 230 in Jasanoff, S. (1990) *The fifth branch: science advisors as policy-makers.* Harvard University Press: Cambridge, MA.

interactions would be to invite open consultation across society about what dimensions of risk actually matter to the public, to invite experts to assess and contribute what is known about the risks of different levels of climate change, and to require politicians and policy makers to argue and negotiate in public about what level of risk is intolerable and to set policy accordingly. This co-production model is much more sympathetic to framing knowledge in terms of risk, in which uncertainties are inherent and visible. It also recognises that expert judgement, invoking Bayesian beliefs about the credibility of knowledge, is integral to shaping the scientific knowledge that can contribute to this collective process of policy making.

These three different ways of using science for policy can co-exist within the same nation at the same time. Different actors, agencies and institutions may operate under different models. The relationship between science and policy undoubtedly changes over time and is certainly different between different nations. For example, the role of science in American political culture is quite different from its role in the UK or Europe, which is different again from its role in China. This has implications for climate change science. Even if the IPCC were publishing assessments of universally agreed and finally definitive knowledge about climate change – and we have shown that there are limits to such universalist claims – the way in which this knowledge would shape policy debates and policy formation in different countries, and the authority carried by scientific statements about climate change in different cultures, would differ.

3.6 Summary

So why do we disagree about climate change? Many of the disagreements that we observe are not really disputes about the evidence upon which our scientific knowledge of climate change is founded. We don't disagree about the physical theory of the absorption of greenhouse gases demonstrated by John Tyndall, about the thermometer readings first collected together from around the world by Guy Callendar, or about the possibility of non-linear instabilities in the oceans articulated by Wally Broecker. We disagree about science because we have different understandings of the relationship of scientific evidence to other things: to what we may regard as ultimate 'truth', to the ways in which we relate uncertainty to risk, and to what we believe to be the legitimate role of knowledge in policy making. To help our discussions about climate change and to allow scientific knowledge to play its part in such discussions, we need to recognise the limits to scientific knowledge. And we need to appreciate that such knowledge can be (perhaps inevitably will be) transformed in the process of leaving the laboratory and entering the social world.

There are three limits to science that we must recognise. First, scientific knowledge about climate change will always be incomplete, and it will always be uncertain. Science always speaks with a conditional voice, or at least good science always does. Belief in the power of science requires a simultaneous doubt about the final and ultimate adequacy of any scientific knowledge claim. We must recognise that uncertainty and humility should always be essential features of any public policy debate which involves science, not least climate change. Certainty is the anomalous condition for humanity, not uncertainty.

Second, we must recognise that beyond such 'normal' scientific uncertainty, knowledge as a public commodity will always have been shaped to some degree by the processes by which it emerges into the social world and through which it subsequently circulates. What will in the end count as scientific knowledge for public decision making is not necessarily the same knowledge that first emerged in the laboratory. In the production, or better still the co-production, of climate change knowledge for public policy, trust in the processes of science and participation in the social processes of co-production are essential. Without trust and/or participation, scientific knowledge about climate change is unlikely to prove robust enough to be put to good use. The separation of knowledge about climate change from the politics of climate change – a process that has been described as 'purification' – is no longer possible, even if it ever was. The more widely this is recognised the better.

Third, we must be more honest and transparent about what science can tell us and what it can't. We should not hide behind science when difficult ethical choices are called for. We must not always defer to 'science' or to the 'voices of scientists' when we need to make decisions about what to do. These are decisions that in relation to climate change will always entail judgements beyond the reach of science.

All of the above will help us to conduct more honest and open debates about what climate change means for us and what we should do about it. It means that if we *do* disagree about the science of climate change then we can base these disagreements around legitimate causes for difference, rather than on spurious ones. But this will come nowhere near fully exhausting the reasons we disagree about climate change. On occasions we will appear to disagree about the science of climate change when really we are disagreeing about things that have little, if anything, to do with science. There are many more dimensions to consider than just the performance of science. In the next chapter we explore another one of these: the different ways in which we value things, both material and non-material, and how we bring these different ways of valuing into our individual and collective decision making.

FURTHER READING FOR CHAPTER 3

Collins, H. and Evans, R. (2007) **Rethinking expertise**. University of Chicago Press.
This book answers the question 'what does it mean to be an expert?' While it is probably more suited for those who want to explore the nature of scientific knowledge at a more advanced level, its opening chapter helpfully outlines the dilemmas we face in society about drawing lines between expert and non-expert knowledge. The book also reflects

on the nature of scientific authority and how we distinguish 'good' science from 'junk' science.

Funtowicz, S.O. and Ravetz, J.R. (1993) Science for a post-normal age. **Futures 25,** 739–55.
This short article is the best and earliest description of the idea of post-normal science, which draws its inspiration from ideas about the changing role of science in a risk society where knowledge is uncertain and yet there is a demand for policy intervention. Although not developed specifically with climate change in mind, climate change science fits their description of 'post-normal' very well.

Miller, C.A. and Edwards, P.N. (eds) (2001) **Changing the atmosphere: expert knowledge and environmental governance**. MIT Press: Cambridge, MA.
This edited volume presents a series of detailed empirical studies of the relationship between climate change science and the ways in which climate change policies are developed, debated and implemented. Using perspectives from history, sociology and philosophy, the chapters cover issues such as the funding of climate change science, the role of experts, and the formation and governance of the IPCC.

Millstone, E. (2005) Analysing the role of science in public policy-making. Chapter 2 in van Zwanenberg, P. and Millstone, E. (eds), **BSE: risk, science and governance**. Oxford University Press.
This book chapter reviews the different ways in which the relationship between science and policy has been conceived and the implications of each viewpoint. A range of different decision models are outlined, each of which has different implications for science and its role in influencing policy.

Pielke, R. Jr (2007) **The honest broker: making sense of science in policy and politics**. Cambridge University Press.
This concise and accessible book is a good general introduction to the ways in which science can be used in policy. Pielke suggests there are four such roles: with scientists acting as a pure scientist, as a science arbiter, as an issue advocate or as an honest broker. His personal view is that the last is the most appropriate, but his call is for scientists to be clearer in their own minds and in their interactions with others about which role(s) they are adopting.

FOUR

The Endowment of Value

4.1 Introduction

Within months of George W. Bush becoming President of the United States of America in January 2001 it was made very clear that economics was at the heart of arguments about climate change policy. Writing to Republican Senator Chuck Hagel and colleagues a few months later, Bush explained: 'As you know, I oppose the Kyoto Protocol because it ... would cause serious harm to the US economy. The Senate's vote, 95–0, shows that there is a clear consensus that the Kyoto Protocol is an unfair and ineffective means of addressing global climate change concerns.'[1] And later that spring, in a White House press statement, he said: 'For America, complying with those [Kyoto] mandates would have a negative economic impact, with layoffs of workers and price increases for consumers. And when you evaluate all these flaws, most reasonable people will understand that it's not sound public policy.'[2]

Bush's claimed reason for withdrawing the USA from the Kyoto Protocol was – at least to domestic audiences – because of its perceived

---

[1] 13 March 2001 letter from George W. Bush to senators Hagel, Helms, Craig and Roberts.
[2] 11 June 2001 White House press release from the President.

damage to the US economy and workforce rather than because the science of climate change was uncertain or incomplete.

A different type of argument about climate change, but again one in which the language and analysis of economics was central, was highlighted a few years later in 2004 when the Copenhagen Consensus Centre in Denmark released the results of an exercise to set priorities for confronting some of the world's greatest challenges. Convened and led by the Danish statistician Bjørn Lomborg, and sponsored by the international magazine *The Economist*, this exercise consisted of a panel of eight distinguished Nobel Laureate economists answering the question: 'What would be the best ways of advancing global welfare … supposing that an additional $50 billion of resources were at governments' disposal?'

Guided primarily by considerations of economic costs and benefits, the panel ranked seventeen policy measures in terms of their effectiveness in advancing human welfare.[3] Control of HIV/AIDS and addressing hunger through the provision of micronutrients were the two measures most favoured by the panel. But when the results of the exercise were publicly released in May 2004, much of the media and policy commentaries focused on the three policies that were at the bottom of the list: the climate change policy measures of the Kyoto Protocol and two variants of a carbon tax. The Copenhagen Consensus conceded that global climate change was an important world challenge and urged more research into affordable carbon-abatement technologies. But within the economic framework used by the panellists, the pay-back to human welfare of these proposed climate policy measures was judged to be insufficient to allocate them any of their (hypothetical) $50 billion.

The refusal of the USA to ratify the Kyoto Protocol – and for similar economic reasons Australia also, until late in 2007 – and the

---

[3] Lomborg, B. (ed.) (2004) *Global crises, global solutions*. Cambridge University Press. The headline results from the Consensus were released in Copenhagen on 29 May 2004.

vociferous arguments about the legitimacy of the economic analyses used by the Copenhagen economists, illustrate two important things about economics and climate change. First, economic arguments carry considerable weight in political decisions about climate change and, second, economic analyses can yield seemingly very different evidence with which to inform those decisions. The economic analyses used by George Bush to justify withdrawal of his country from the Protocol were framed and interpreted quite differently than were those used by the Member States of the European Union, for whom implementing the Protocol became an integral component of economic policy in the region. And, two years after the Copenhagen Consensus argued against early intervention on climate change, the Stern Review on the economics of climate change, published by the UK Government in October 2006, reached a diametrically opposite conclusion: the (large) costs of inaction on climate change greatly outweighed the (modest) costs of taking early initiatives to slow down the rate of warming.[4]

As we saw in Chapter 3: *The Performance of Science*, conventional expectations of science being able to discover the objective 'truth' of how the world works and to make reliable predictions of the future are challenged by the complexities of climate change. The expectations of economics – a discipline described famously by Thomas Carlyle in the nineteenth century as 'the dismal science' – have perhaps never been as lofty. There is certainly a longer list of forecasts going awry in economics than in most areas of scientific prediction. And yet economics remains both essential and powerful as a way of exploring and informing the sorts of decisions we have to take, for example as individuals about our consumer choices, by societies about investments in health care, or globally in relation to climate change. But if the

[4] Stern Review (2006) *The economics of climate change: the Stern Review.* Cambridge University Press. The Stern Review was released in London on 30 October 2006.

unified practice of science has been fractured by the recent demands of 'post-normal' environmental issues, there is an even greater multitude of ways in which economic analyses can be performed.

Individuals and societies ascribe value to activities, assets, constructs and resources in many different ways. **One of the reasons we disagree about climate change is because we ascribe these values differently.** This chapter explores some of the different economic frameworks used in climate change debates and explains why the choice of valuation framework is so important when deciding what to do about climate change. Some of the central concepts in economics as they relate to climate change are explored in Section 4.2; for example, gross domestic product as an index of prosperity, cost–benefit analysis and its application in welfare economics, and the importance of the discount rate in all economic analyses.

The most widely cited economic assessment of climate change in recent years – the Stern Review – uses many of these concepts in articulating a powerful economic case for strong global climate policies. In Section 4.3 we examine the framework, assumptions and value judgements made in the Stern Review and then present two sets of arguments which have claimed that the framework used by the Stern Review was, respectively, too conservative or too radical. This comparative critique reveals that many of the economic reasons for disagreement about what to do about climate change are rooted in different choices of analytical frameworks, different ethical positions about our responsibility for the future world, and different attitudes to risk. These various dimensions of how and why we attach value to things are elaborated in Section 4.4. Radically different prognoses for addressing climate change emerge, depending on what value system is adopted.

4.2 Some Applications of Economics

Economics is the study of how people choose to use and allocate resources. Resources include the time and talents people possess, natural resources such as land, air and water, the built infrastructure, and the knowledge of how to combine these resources to create useful products and services. Professional economists recognise a number of different ways of analysing the production, allocation and use of such products and services. The dominant framework continues to be that of neoclassical welfare economics, which is strongly linked to a view of the individual as a rational consumer and of the market as the dominant means by which values are revealed. Other frameworks include Marxian economics, new institutional economics, and ecological economics. In the context of climate change, Marxian economists would focus on differential access to power and resources between rich and poor; the latter being the powerless victims of climate change. The new institutional economics framework would examine the institutional reasons why climate change has occurred, focusing on transaction costs, organisational hierarchies, modes of governance, and social capital. Ecological economists would work with the full range of environmental goods and services which are offered by climate and which are threatened by climate change. This would include those goods and services for which there is currently no market and for which ecological economists would need to find ways of incorporating their value into the analysis.

None of these frameworks are particularly well suited to analysing the economics of climate change. Whichever framework is chosen will encounter significant difficulties when it is applied to the many novel economic questions raised by climate change. These may be questions such as: What are the costs of limiting carbon dioxide concentrations in the atmosphere to 450 ppm? What are the benefits of keeping global warming to 2°C rather than 3°C above the pre-industrial temperature? What would be the effect on Europe's economy of introducing a carbon tax of, say, €50 per tonne of carbon? To understand the disputed nature of the economic answers to questions such as these, we need to understand a few key economic concepts as they apply to climate change. How do we measure welfare? How do we compare the costs and benefits of climate change policies? How should we weight the future in such comparisons? And how do we do economics when we don't know for sure what the risks associated with climate change will be?

Measures of welfare

The most ubiquitous index for measuring economic performance is gross domestic product (GDP). A nation's GDP – defined as the total market value of all final goods and services produced within a country in a given period of time – is one of the ways of measuring the size of its economy. But GDP is limited to goods and services that have a market value. As an overall indicator of wealth, of human and ecological well-being, GDP is completely inadequate. Using GDP as the sole measure of economic growth, and therefore implicitly of social 'progress', introduces distortions into public policy. This has been commented on by American economist Herman Daly, one of the founders of ecological economics in the 1970s: 'Current economic growth [measured by GDP] has uncoupled itself from the world and has become irrelevant. Worse, it has become a blind guide.'[5] It was also the key insight of the report of the 1987 Brundtland Commission, *Our Common Future*, which we examine in greater depth in Chapter 8: *The Challenges of Development*.

When considering the economics of climate change, it is frequently necessary to consider costs and benefits which are not captured by GDP, to include goods and services which matter (greatly) to us but which do not have a market price. Crucially, we need ways of attaching value in our analysis to things that are offered to us by climate, and which might therefore be diminished or enhanced by climate change, but which do not appear as goods in the market-place or enter into the flow of products captured by GDP.

Two of the most important such categories are natural capital and aesthetic values. Natural capital comprises the renewable and

---

[5] p. 945 in Daly, H. E. (1972) The economics of zero growth. *American Journal of Agricultural Economics* 54(5), 945–54.

non-renewable goods and services offered by ecosystems, while aesthetic values would include those deriving from landscapes and skyscapes, sacred places and spiritual meaning (see Box 4.2 for examples). Whichever framework is used, an economic analysis of climate change that *excludes* considerations of natural capital and the aesthetics of climate – or assumes that the loss of such values can be compensated through monetary transfers or payments – will offer a very different prognosis than an analysis which *includes* such considerations. We will see later in this chapter how crucial a difference this may make.

Cost–benefit analysis

The most widely applied framework for exploring the economics of climate policy is cost–benefit analysis. This framework allows the costs of reducing emissions of greenhouse gases into the atmosphere to be compared with the benefits of doing so. The benefits of reducing greenhouse gas emissions are usually estimated to be the damage avoided by reducing the magnitude of climate change. To be used in a comparative analysis the damage due to climate change and the costs of mitigation have both to be expressed in monetary terms, usually as a dollar (or euro) value per ton (or tonne) of carbon dioxide either emitted or averted – what is called the 'social cost of carbon'. Of these two quantities, calculating the damage avoided is much harder than calculating the costs of reducing greenhouse gas emissions, for the reasons outlined in Box 4.1.

The economic principle behind cost–benefit analysis is easy to state but, in the case of climate change, difficult to implement. In principle, if it costs less to avoid emitting one tonne of carbon dioxide[6] than the

---

[6] Of course, it need not literally be carbon dioxide; other greenhouse gases, such as methane or nitrous oxide, could be used in the analysis, but for analytical purposes all costs are usually indexed against carbon dioxide, or the equivalent amount of carbon dioxide that would yield the same amount of damage as a unit of any other greenhouse gas.

damage caused by adding that tonne to the atmosphere, then policy efforts should be directed towards avoiding that emission. And such policy efforts should continue until such time as the cost of avoidance starts exceeding the 'benefit' that would accrue from the reduction in climate change thus brought about. In other words, it makes sense to keep reducing emissions for as long as the benefit of doing so exceeds the cost. Cost–benefit analysis seeks to optimise policy actions using explicit monetary criteria and is widely applied across a range of private and public policy issues; for example, in deciding on large-scale building or transport projects such as a city metro system.

However, many of the characteristics of climate change challenge the standard application of cost–benefit analysis as an economic tool to support decision making. Because changes in climate affect all regions, the costs and benefits have to be worked out at a global scale; and this is rarely achievable in any policy context. Because the potential damage caused by climate change includes many things that do not have a market value, estimating the benefits of avoided greenhouse gas emissions and converting them into monetary units is very difficult. Because we are uncertain about many of the risks that climate change may cause, it is very hard to put numbers on the consequences of these risks, even for monetised assets. For example, estimates of the social cost of carbon range from $0 per tonne of carbon (tC) to over $2,000/tC (see Box 4.1). And, finally, because climate change is a long-lived phenomenon – operating over decades, generations and centuries – how we value the distant future becomes an essential, if not *the* essential, component for cost–benefit analysis applied to climate change.

For all these reasons – a global-scale phenomenon affecting the distant future and with uncertain consequences, many of which have no market value – the application of conventional cost–benefit analysis to climate change policy making becomes at best very difficult and at worst impossible. Cost–benefit analysis might be very effective as an economic guide to relatively well-contained decisions affecting, at most, two or three decades into the future and where costs and benefits can be reasonably well estimated. But there are many who see it as an inappropriate tool to use in the case of climate change. Instead, it acts as a further locus for disagreement about climate change.

Box 4.1: The Social Cost of Carbon

To undertake a cost–benefit analysis of climate change, it is necessary to estimate the worldwide incremental damage that would be caused by emitting one tonne of carbon dioxide (or the equivalent of other greenhouse gases) at a point in time and to express this damage in monetary terms. This is what is meant by the 'social cost of carbon'. Alternatively, it can be expressed as the marginal benefit, or avoided damage, resulting from averting the emission of a tonne of carbon into the atmosphere, and in cost–benefit analysis can be set against the cost of such aversion.

Estimating the social cost of carbon is not straightforward and calls into play a number of significant assumptions and value judgements about the future. At its simplest we can express the elements that contribute to the social cost of carbon (SCC) as follows:

$SCC \approx f\,(damage,\ adaptive\ capacity,\ discount\ rate,\ equity\ weighting)$

Each of these elements is problematic. Estimating the potential damage caused by climate change and expressing such damage in monetary terms itself entails at least three separate activities. Thus the damage term can be expressed as follows:

$Damage \approx f\,(market\ costs,\ non\text{-}market\ costs,\ risk\ of\ abrupt\ climate\ change)$

The net market costs[7] of different magnitudes of climate change have to be aggregated across the world and for all economic sectors. Some sectors are easier to analyse than others. For example, it is relatively easy to estimate the changes in heating and cooling demand in buildings in different regions under different magnitudes of warming and then to convert these into an estimate of the net economic cost of meeting these changes in energy demand. On the other hand, it is very difficult to estimate the changes in tourist activity in different parts of the world that might occur in response to climate change and to make a guess at the gains or losses in economic activity that would ensue in each region.

For aspects of our lives that we value and which are associated with climate, but for which there is no market exchange, the problems multiply. For example, if a breeding site for a rare migratory bird species was lost in one country because of sea-level rise, but was then re-established in another country, how does one evaluate the monetary cost of this change in location? Does a loss of wildlife in one country offset the gain somewhere else? Or if snowy winters in a mid-latitude country no longer occurred because of climate change, what would be the (lost) monetary value of the experience of children playing in the snow? And with respect to the damage that might be caused by a major disruption to some feature of the climate system – such as an irreversible change in the south Asian monsoon – how does one incorporate the (presumably) large but unknown social costs associated with such a (presumably) unlikely outcome?

But estimating the (monetised) net damage costs of climate change is only one part of the problem of estimating the social

---

[7] It is important to note that it is 'net' market costs that matter here, i.e. the loss of market value minus any gains in market exchanges which may also occur from climate change. Climate change is not all about costs, so market benefits must be estimated as well. For example, as climates change, new agricultural activities and produce may become viable at the current northern thermal limits.

cost of carbon (see above equation). The importance of adaptation to changes in climate has also to be considered. There is obviously the potential for some of the 'damages' alluded to above to be avoided through planned or reactive adaptation. Individual companies, organisations, communities and societies are responsive and innovative entities and will not be passive in the face of climate change. Adaptation has the potential to reduce the apparent costs of climate change, or even to yield benefits, although some adaptations will themselves incur costs. The foresight to be able to estimate all of these second- or third-order adjustments of society to climate change is rarely found.

The last two elements of the social cost of carbon introduce very important ethical considerations: What weight do we give to future generations (the discount rate) and how do we attach weights to the lives of individuals who are richer or poorer than ourselves (equity weighting)? The discount rate is discussed in the main text below, but equity weighting is equally important. For example, when converting estimates of death or disease caused by climate change into monetary units, how is the value of a Chinese life treated in comparison with a Norwegian life? Traditional neoclassical economists would attach different monetary values to these two lives on the basis of revealed costs,[8] whereas a different set of ethical considerations might argue that they should be weighted equally.

---

[8] The controversy about the 'value of a statistical life' in the context of climate change economics was particularly heated in the 1996 Second Assessment Report of the IPCC. Using conventional neoclassical economic procedures it was suggested that the value of 'statistical life' in the developed world (~$5m) might be at least ten times higher than in the developing world (~$0.5m). See Fankhauser, S., Tol, R. S. J. and Pearce, D. W. (1998) Extensions and alternatives to climate change impact valuation: on the critique of IPCC Working Group III's impact estimates. *Environment and Development Economics* 3, 59–81 for a further discussion about this controversy.

> The above discussion has hinted at the difficulties of estimating the social cost of carbon and explained why different analysts in different countries, following different assumptions and ethical principles, will arrive at different estimates. It is not surprising, then, that a review of the published estimates for the social cost of carbon should show a range from $0/tC to over $2,000/tC, with a mean estimate at about $90/tC.[9] Even within one country there can be major disagreements about the appropriate social cost of carbon. In the UK, for example, there has been a long-running argument about whether the social cost of carbon to be used in government policy should be closer to $14/tC or $140/tC, a full order of magnitude difference. Clearly, if we cannot agree about the potential damage caused by future climate change then it makes it much harder to find agreement about the overall goals and urgency of climate change policy.

Discounting the future

At the heart of all cost–benefit analyses which inform decisions which have consequences that last for more than a few years is the selection of an appropriate social discount rate. For climate change, whose effects might be felt for several generations, if not centuries, the choice of discount rate has a very large impact on the result of any cost–benefit analysis and, consequently, on any attempt to apply economic analyses to climate change decision making.

Questions about social discount rates in economics are questions about the relative importance in societal decisions of the welfare of future generations relative to that of the current generation.

---

[9] See *Stern Review*, p. 323. It should also be noted that these costs can also be cited as a cost per tonne of carbon dioxide, rather than per tonne of carbon, in which case the value has to be divided by 3.7.

The choice of discount rate – more accurately the 'rate of pure-time preference' or time discount rate – in any analysis is usually based on either descriptive (empirical) considerations or prescriptive (ethical) considerations. Descriptive approaches tend to lead to higher discount rates (for example, 4 per cent per annum or more), while prescriptive approaches usually suggest lower rates (less than 3 per cent or even, in some cases, zero). In the context of climate change, the debate about discount rates arises because the benefits (i.e. avoided damages) of today's investments in mitigation activities (i.e. that lead to reductions in greenhouse gas emissions) will only be realised by future generations.[10] Choosing a discount rate – either choosing between an empirically observed rate and a prescriptive one or, if a prescriptive rate, which one – requires ethical judgements to be made by the analyst about the relative importance of future generations as opposed to the present generation.

The argument *for* a low discount rate is essentially an ethical one: the welfare of a future person should matter just as much to us as the welfare of a living person. There should be no discrimination against the unborn just because they are not yet born. In the context of climate change the argument runs thus. Any damages from climate change will increasingly be borne by future generations. If those people were able to speak to us today, they would ask us to forgo some of our own consumption so that they could enjoy the future benefits of a reduced change in climate. In other words they would ask us to divert some investment today into climate change mitigation, thereby reducing, by some amount, future damage from climate change. This

---

[10] Because the climate system responds only very slowly to changes in the rate of greenhouse gas emissions, physical climate displays considerable inertia with respect to such changes. Current understanding suggests that anthropogenic climate change over the next forty years or so will largely be a response to emissions of greenhouse gases from the past. In other words, our achievements in reducing emissions today and in the future will only slow down the rate of climate change beyond about 2050, and not before.

prescriptive approach to discount rates is the one used in the analysis adopted by the Stern Review (see Section 4.3).

The argument *against* a very low discount rate is twofold. Empirically, our consumption habits suggest that we *do* give lesser weight to the welfare of future generations than to our own. Our spending patterns suggest that we prefer consumption today over deferred benefits tomorrow, whether these deferred benefits are for us or our descendants. The second part of the argument against a low discount rate derives from the presumption that future generations are likely to be much richer than we are today.[11] If this is so, then they will have greater spending power than we do in order to cope with, or adapt to, any future consequences of climate change. In this case, the interests of future generations should not be favourably weighted (i.e. through a low discount rate). Instead of making investments today in mitigating future climate change, a better use of that investment would be in reducing the vulnerability to climate risks of the poor of today's generation. As British economist David Pearce explains, a low discount rate might lead to a 'sacrifice of resources today … at the expense of transfers of income to poor people today. If so, the poor today may bear sacrifices in terms of forgone benefits in order to benefit their richer descendants'.[12] Or, to put it more bluntly, we should not feel too obliged to make sacrifices today so as to make future generations even richer than they might be anyway; any more than, say, our parents should have given up their *only* car so that we would be in a position to buy a *second* car for our household.

There are arguments to be made in favour of both of these positions. The choice of discount rate has in the end to be based on a

---

[11] Standard economic analyses nearly always assume a growth in real terms of the economy, for example an annual growth in per capita income often of between 1 and 2 per cent. The former rate would double individual wealth within 70 years, the latter rate within 35 years. The Stern Review assumed a 1.3 per cent growth rate, with wealth doubling every fifty-four years.

[12] p. 363 in Pearce, D. (2003) The social cost of carbon and its policy implications. *Oxford Review of Economic Policy* 19(3), 362–84.

much wider set of considerations than just economic ones. Our attitude to future generations is important along with how wealthy we think they will be. Also central to our reasoning has to be whether we think that some of the consequences of climate change will be so great that either they cannot be reversed or adapted to, or in some other way cause the loss of aspects of natural capital or aesthetic value that cannot be substituted through monetary transfers. Which brings us to consider our attitudes to risk, danger and precaution – and how these attitudes differ.

Uncertainty, risk and precaution

We perceive and evaluate risks in different ways, often basing our individual or collective decisions on quite different attitudes to risk. These might be quite mundane decisions – like whether to take out an insurance policy to cover us against a new cooker malfunctioning – or, more significantly, how stringent to make airport security in the face of possible terrorist attacks. At its most obvious, there are those of us who are risk-takers and there are those who are risk-averse. We will explore these attitudes more thoroughly in Chapter 6: *The Things We Fear*.

It is when we consider the likelihood and severity of future climate risks, and our belief in these likelihoods, that we begin to see a further limitation of conventional economic analyses of climate change. We have seen above how our evaluation of uncertainty in relation to future large climate risks affects estimates of the social cost of carbon (Box 4.1) and also how ethical considerations are central to the choice of social discount rate. If these risks seem very large to us, and if they suggest changes in climate that we believe no future generation can adapt to, then the results of any formal cost–benefit analysis might be trumped by the application of the precautionary principle.

The precautionary principle is summarised in the aphorism, 'it's better to be safe than sorry'. It states that if an action *might* conceivably cause severe or irreversible harm then, in the absence of a scientific consensus that harm would *not* ensue, the burden of proof falls on those who would advocate taking the action. Applied to climate change it means that since there is *prima facie* evidence that a change in climate would induce some harm (maybe a lot of harm), then those who would resist efforts to reduce the growth in greenhouse gas emissions have to demonstrate that those avoidable emissions are harmless.

As we will see below in the debate about the Stern Review, some of the differences of interpretation about economic analyses of climate change have nothing to do with economics at all. It is not about whether a cost–benefit analysis, with all of the assumptions and judgements that lie behind the numbers, tells us that we should act, whether quickly or slowly, and at what cost. It is actually about whether or not we judge the (largely unknown) risks associated with climate change to be so potentially large and undesirable that we distrust *any* cost–benefit analysis, whatever the assumptions. In this case, some would determine that the precautionary principle demands that we must immediately do all that we can to mitigate climate change. Invoking the precautionary principle takes climate change well beyond the realm of conventional economics and places it firmly in the court of ethics and risk assessment. We examine further these two dimensions of ethics and risk in, respectively, Chapter 5: *The Things We Believe* and Chapter 6: *The Things We Fear*.

4.3 The Stern Review and its Critics

We are now in a position to consider the economic arguments contained in the Stern Review in more detail. More importantly, we need to understand why the conclusions of the Stern Review were criticised by other economists and by climate change commentators for being either too radical or too conservative. Those who thought that an authoritative economic analysis of climate change would finesse the uncertainties of the science of climate change and define some obvious and agreed courses of political and social action were wrong.

The essence of the Stern Review

The Stern Review on the economics of climate change was commissioned by the Treasury – the Finance Ministry – of the UK Government, and the review team undertook its work during 2005 and 2006. It was chaired by a former World Bank economist, Nicholas Stern, from whom it took its name. At its launch in London on 30 October 2006, the British Prime Minister, Tony Blair, claimed: 'This is the most important report on the future published by the Government in our time in office … what the Stern Review shows is how the economic benefits of strong early action easily outweigh any costs. It proves tackling climate change is the pro-growth strategy.'[13]

The headline sound-bites were certainly arresting – 'Economists warn climate change will cost world trillions if governments fail to act', 'We have been warned: now everyone should understand why we have to combat climate change', 'Warming called [a] threat to [the] global economy'.[14] The key recommendation of the Stern Review, arrived at from economic reasoning, was for urgent, immediate and sharp reductions in greenhouse gas emissions, with the aim of stabilising carbon dioxide equivalent concentrations in the atmosphere at between 450 and 550 ppm.

To reach its conclusion that the costs of inaction on climate change greatly outweighed the costs of taking early and urgent action, the Stern Review used a number of different economic concepts and frameworks in its work. Although starting from a general cost–benefit analysis framework, the Review paid considerably more attention to questions of uncertainty, risk and equity than is usual in conventional cost–benefit analysis. By claiming that 'climate change is the greatest market failure the world has ever seen', the Review clearly adopted a fairly conventional welfare economics perspective. On the other

[13] Speech by Tony Blair, 30 October 2006, London.
[14] Headlines from Friends of the Earth press release, 13 October 2006; *New Scientist* editorial, 4 November 2006; *Washington Post*, 31 October 2006.

hand, some of its assumptions and ethical judgements went beyond conventional economics. A very low time discount rate of effectively zero,[15] combined with a per capita growth rate of 1.3 per cent, yielded an overall social discount rate of 1.4 per cent, much lower than the more conventional range of 3–6 per cent. The Review also adopted a fairly high estimate of $300/tC for the social cost of carbon, again based on a wider set of considerations than conventional welfare metrics.

Since being published, the Stern Review has had great visibility in debates about climate change policy. Despite being commissioned by the Finance Ministry of the British Government rather than working under a mandate from the United Nations, the Review seems at times to have acquired almost as much authority as an IPCC report. It formed a major plank of the British Government's climate change diplomacy during 2007 and beyond, and has provided an easily cited source for arguments in favour of strong and early action on climate change mitigation. But its economic diagnosis of climate change and its prognosis for climate policy have not gone unchallenged. There are other ways of approaching the economics of climate change; ways which would suggest either a more conservative prescription for action or a more radical one. We will first examine the counter-arguments from the more conservative end of the spectrum.

More conservative analyses

One set of criticisms of the Stern Review operates within the same basic framework of cost–benefit analysis, but challenges the assumptions used in the Review about discount rates, damage costs and the risks of catastrophic events. In essence, these critics question the way in which the Review arrived at the headline result that the costs of unmitigated climate change will be 'between 5 and 20 per cent of global

---

[15] In fact 0.1 per cent was used rather than zero on the outside chance that the human species becomes extinct within the next two centuries.

GDP each year, now and forever'. While not necessarily disagreeing with the Review that an economic analysis would show that action needs to be taken on climate change, these critics have suggested that rather less drastic or immediate interventions were warranted.

Exploring this position helps us understand why economists disagree about action on climate change. The various arguments reduce, in essence, to three: What is the appropriate discount rate, what are the damage costs of climate change, and how likely are the most catastrophic outcomes of climate change? Each of these questions revolves around matters of judgement – judgements which take us beyond observable or predictable realities, and about which science is therefore either silent or deeply uncertain.

Veteran American economist William Nordhaus has been most vocal in challenging the choice of the very low discount rate used in the Review. Rather than taking an 'extreme prescriptive' approach to the choice of discount rate, Nordhaus favours an approach that, while still incorporating ethical judgements, would be more closely tied to the empirical behaviour of markets and individual economic behaviour. His advocated approach would lead to the damage of unabated climate change being estimated at between only 0.5 and 2 per cent per annum of global GDP, a full order of magnitude lower than that derived by the Stern Review.

The root of the disagreement concerns the importance of the welfare of future generations relative to that of our own generation. If we think that all generations that come after us should count as equally important in our decision making as our own generation, taking little heed of whether those future generations are richer or poorer than us, then climate change looms as a more urgent problem to invest against today. We should apply a low or zero discount rate: the position of the Stern Review. On the other hand, if (as Nordhaus and some other economists would argue) the way we act and make decisions today in other areas of economic life reveals that we don't weigh the interests of all future generations equally against our own, then we should apply a higher discount rate. In this case, the future damages of climate change weigh less heavily on our conscience as we make decisions today about investment options. Neither science, nor the principles of economics, can arbitrate between these two positions. Wider sets of ethical and moral reasoning need to enter the debate, as we will see in Chapter 5: *The Things We Believe*.

If the appropriate discount rate to use in the analysis of climate change is unresolved by economics, then the physical (and hence economic) damage caused by climate change, and the risks of catastrophic changes, are poorly understood by science. We rehearsed some of the reasons for this earlier when discussing the social cost of carbon (Box 4.1). By adopting a social cost of carbon of $310/tC, the Stern Review opted for an estimate towards the higher end of the published range. Other economists, using the same welfare economics analytical framework as Stern, have argued that this estimate is too high, and hence the urgency of the problem as painted by the Review has been exaggerated. The issues here are really about the interpretation of scientific uncertainties and about the extent to which future generations will be able to adapt successfully to some of the forthcoming changes in physical climate.

We can illustrate the issues involved – and the reasons for disagreement – by exploring the way in which 'economic damage functions' are constructed by economists. A damage function is normally expressed as the loss of global world (economic) production (GWP) for each degree Celsius of global-average warming. The scientific literature on this function is not large, but what research has been completed suggests that, up to about 2.5°C of warming, about 1 per cent (between zero and 2 per cent) of GWP might be lost. Beyond that temperature, assumptions have to be made about the shape of the damage curve. Does it go on rising in a straight line (e.g. 2 per cent loss at 5°C, 3 per cent loss at 7.5°C, etc.), or does the curve start steepening? If the damage curve is non-linear then exactly how non-linear – does it start rising steeply or not so steeply (see Figure 4.1)?

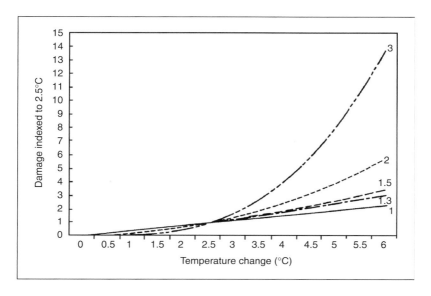

FIGURE 4.1: *Some possible climate change damage functions expressed as global economic damage indexed against damage estimated to result from a global-average warming of 2.5°C (i.e. index value = 1). The different curves reflect different choices about the value of the damage function exponent (value shown for each curve; e.g. 1.3%).*
Source: *Dietz et al. (2007).*

A further question concerning such damage functions is how do they represent economic damages associated with possible, but unlikely, 'catastrophic' climate impacts (for example the collapse of the West Antarctic Ice Sheet)? There are no scientific studies available on which to base such assumptions. Indeed, it is not at all obvious how one would even design a scientific study using observations and/or theory to answer such a question.

The result of these ambiguities is that the shape of damage functions in economic analyses is largely assumed rather than based on any empirical or theoretical evidence. Different economic analysts use different shaped curves, hence different economic analyses yield different diagnoses of the severity of climate change. Using the same ethical principles and analytical framework as the Stern Review, but by altering the nature of the damage function curve, another economic analysis could yield a very different treatment of climate change. For example, other cost–benefit type analyses by economists such as Nordhaus and the Dutch economist Richard Tol estimate a social cost of carbon of between $40 and $120/tC – well below the $310 used by the Stern Review.

We have demonstrated in this section how different economic analyses using conventional welfare economics approaches can yield quite different economic diagnoses of the climate change problem. By using more conservative assumptions than were made in the Stern Review, climate change would not lead to the future loss of between 5 and 20 per cent of global GDP, as claimed. Yet the economic arguments about climate change are much more diverse than just those examined here. We next explore the equally value-laden perspective of the ecological economists; economists who would reject cost–benefit analysis entirely – even the modified version used in the Stern Review. Instead, they would favour a more radical precautionary approach to designing global climate policy.

More radical analyses

In contrast to the mainstream welfare economists who felt uncomfortable with the conclusions of the Stern Review, other voices adopting the philosophy of ecological economics were also very critical of the analysis presented in the Stern Review. Rather than focusing debate on the appropriate values of the key parameters in the Review – the discount rate, the social cost of carbon – these voices challenged the entire welfare economics framework upon which the Stern Review was predicated. At the heart of this critique is the view that such analytical frameworks cannot address the fundamental values at stake with climate change; a view expressed by ecological economist Clive Spash: 'The [Stern] authors maintain allegiance to an economic orthodoxy which perpetuates the dominant political myth that traditional economic growth can be both sustained and answer all our problems.'[16]

In rejecting welfare economics as the basis for decision making with respect to climate change, ecological economists will often point to the 'irreversible and non-substitutable damage to and loss of natural capital'.[17] By this they mean that some of the changes to the world brought about by physical changes in climate – loss of species, coral reefs or glaciers, or the inundation of coastal lands caused by large sea-level rises – cannot be compensated by any growth in consumption or by monetary transfers. Such changes represent absolute and irreversible losses to natural assets or functions to which we attach great value and which are beyond the reach of market exchange. In this view it makes no sense to debate whether such loss represents 1 per cent or 20 per cent of global GDP. Neither does it make sense to enter into nuanced economic arguments about whether such losses could be compensated by economic growth rates of either 1 or 2 per cent per annum, whether discounted at 0.1 per cent or 6 per cent. The basis of their argument lies well beyond the reach of classical economics: 'Climate change, at least above a certain temperature rise, violates fundamental principles of sustainable development, intergenerational stewardship and fairness and therefore violates the inalienable rights of future generations.'[18]

Such economists are taking climate change out of the realm of utilitarian or conventional welfare-based calculation. They are claiming that the grounds for action lie instead in a rights-based approach to valuation, namely the right of future generations to benefit from and appreciate the same basic natural assets and functions as do our own generation. Their criticism of the Stern Review, and other similar economic analyses, is not that it made a number of non-conventional

[16] p. 706 in Spash, C. L. (2007) The economics of climate change impacts à la Stern: novel and nuanced or rhetorically restricted? *Ecological Economics* 63(4), 706–13.

[17] p. 297 in Neumayer, E. (2007) A missed opportunity: the Stern Review on climate change fails to tackle the issue of non-substitutable loss of natural capital. *Global Environmental Change* 17(3–4), 297–301.

[18] *Ibid.*, p. 299.

ethical judgements, but that it did not drive home sufficiently its own logic of presenting the economics of climate change as primarily about ethics, risk and uncertainty. If it had done so, then the debate would revolve about the rights of future generations rather than about how many percentage points will be lost from global GDP. After all, given the Review's assumptions about continued economic growth at 1.3 per cent per year for the next 200 years, it is always possible for people to say that future generations that are two, four or six times wealthier than the present generation will be able to bear the (even quite extreme) losses posited by the Stern analysis.

The perspective of these ecological economists requires climate change to be framed as a problem about the need to take precautionary action to avoid a disaster, rather than as one about how to ensure continued profitable returns on an investment. The view of the radical deep ecologists challenges the ways in which conventional investments are made, businesses are run and markets conduct their affairs. It sits uneasily with the prevailing economic milieu. We return to this critique in Chapter 8: *The Challenges of Development*.

4.4 Revealing Values in Economics

The Stern Review certainly had the effect of heightening debate around the world about whether one should frame climate change in terms of economics and, if so, then how this should be done. Yet it has not provided an uncontested prescription of what needs to be done about climate change, any more than the various assessments of the IPCC have done. Resolving the valuation, ethical and philosophical disagreements about climate change is beyond the reach of physical science and beyond the capability of economics. It requires articulating and debating different belief systems and different social values. We will explore some of these beliefs in the next chapter, but let us now clarify and sharpen the questions about valuation that our journey into economics has uncovered.

First among these is the weight we give to the welfare of future unborn generations as against the welfare of those in our own generation. Or, put even more crudely, how much do we care about our own welfare (read, 'consumption') rather than the welfare of others (read, 'forgone consumption')? There are two fault lines here. The first disagreement is between those who choose to infer these relative weights from our current behaviour (a descriptive approach) and those who argue that such revealed preferences are not consistent with our expressed preferences or ethical norms (a prescriptive approach). This latter group believes that these ethical considerations should take precedence because people tend to express greater preference for equality with their words than through their deeds. The second fault line concerns the weight we place on the welfare of future generations relative to our own. On the one hand there are those who emphasise our obligation not to bequeath future generations climate risks which may diminish their quality of life (or even their ability to survive), while on the other there are those who would place a higher duty on securing the welfare of those living today (including their life chances).

These different judgements are all captured in the two key parameters – the time discount rate and the equity weighting factor – which together largely define the social discount rate used in cost–benefit analysis (see Box 4.1). Judgements which yield a very low discount rate, as in the Stern Review, contrast with the judgements made by President Bush and the Copenhagen Consensus in the examples cited at the start of this chapter. The Nobel Laureates of the Copenhagen Consensus sought to maximise human welfare today and believed that future generations that are richer than ours will generally be able to cope with or adapt to climate change. On the basis of these assumptions, but only these assumptions, it makes economic sense – on both efficiency and equity grounds – to invest massively in physical and human capital now, especially in poor countries, and to divert substantial funds into tackling climate change only at a later date.

A second fundamental dilemma we face is the extent to which we can discover the value of all the things that matter to us and then commensurate them using some universal metric, usually suggested to be monetary value. The fault line here is between those who believe that, at least to a first approximation, such values can be revealed and commensurated, and those who believe that they cannot. Thus we have the contrast between the Stern Review's claim that climate change represents the 'greatest market failure' the world has seen and the claim from the deep ecologists that climate change represents a threat of major proportions to the basic functioning of the planet and the natural capital upon which we depend for survival. In the former case, solutions to climate change are framed precisely in the language of market prices and externalities. This is in direct contrast to the counter-claim of the ecologists that the market *cannot* act as a corrective agency to 'solve' climate change. Economist Frank Ackerman, writing for Friends of the Earth, put this position thus: 'The profundity of human and ecological loss implied in the portraits of climate change, especially at higher temperatures, is only cheapened and diminished by pretending that all of it has a price.' [19]

A problem for both of these positions is how those 'deep values' – what we earlier called aesthetic values and what in other contexts have been referred to as 'sacred values' [20] – are established; values that are not normally expressed through market prices. The Stern Review, by adopting a welfare approach, had to make some heroic assumptions about the monetary losses associated with such things as coral reef decline, reductions in biodiversity, and the outside chance of catastrophic climate events. The deep ecologists also have a difficulty. By adopting their precautionary position, they do not need to express the worth of a songbird

[19] p. 22 in Ackerman, F. (2007) *Debating climate economics: the Stern Review versus its critics*. Report for Friends of the Earth, UK. See www.foe.co.uk/resource/reports/debate_climate_econs.pdf [accessed 30 May 2008].

[20] 'Sacred values differ from material or instrumental ones by incorporating moral beliefs that drive action in ways dissociated from prospects for success.' Examples might be the welfare of one's family and country or one's commitment to religion, honour and justice. See Atran, S., Axelrod, R. and Davis, R. (2007) Sacred barriers to conflict resolution. *Science* 317, 1039–40.

or an ice sheet in monetary terms – in other words, they don't have to justify climate policy using narrow welfare considerations. But they *do* have to deploy some rational argument about the innate spiritual, sacred or survival value of things offered to us by climate.

These values are not only diverse between cultures (see Box 4.2), but they change over time and, for analyses of long-term issues such as climate change, they have to be assumed for generations and cultures yet to come. How the cultural services and functions of physical climate are valued changes from generation to generation; they are unstable both individually and collectively. Wind, for example, was a highly valued energy resource in the international maritime trade before the 1850s, before losing most of its value to the fossil fuel economies of the late twentieth century. Yet it is now regaining new, unimagined value in those economies where wind energy is displacing coal, oil or gas in the generation of electricity. As we saw in Chapter 1: *The Social Meanings of Climate*, not only do physical climates change over time, but so also do the cultural values we attach to climate. Again, heroic assumptions have to be made about how these values will change over the long-term future.

Box 4.2: The Cultural Values of Climate[21]

How do we attach value to our physical climate and which of these values are at risk from climate change? How might climate change offer new opportunities for valued cultural symbolism? Is the only useful way to express such values in monetary terms?

---

[21] Sources for Box 4.2: Adger, W. N., Lorenzoni, I. and O'Brien, K. (eds) (2009) *Adapting to climate change: thresholds, values, governance*. Cambridge University Press; Ford, J. D. (2008) "Where's the ice gone?": climate change vulnerability and adaptation in an Inuit community. Paper presented at the conference '*Living with climate change: are there limits to adaptation?*' Tyndall Centre: London, 7–8 February 2008; and Sanders, T. (2003) (En)Gendering the weather: rainmaking and reproduction in Tanzania. Chapter 5 in Strauss, S. and Orlove, B. (eds), *Weather climate culture*. Berg: Oxford.

Few attempts have been made to explore fully the aesthetic, symbolic and spiritual values that we derive from our climate – what might be called the cultural rather than the physical value of climate (cf. Chapter 1). And yet if we are to truly understand what motivates us to take action, individually or collectively, to restrain climate change we need a more complete understanding of these values which remain beyond the reach of the market. Let us look at some examples.

The Inuit of northern Canada have developed a complex culture and economy that derives much of its character from landscapes, resources and rhythms that are fashioned by a particular climate. Sea-ice acts as an essential platform for hunting walrus, seal and polar bears. The frozen ocean surface also provides an important transportation route, allowing access between islands. Arctic landscapes and the deep seasonality of these regions lend themselves to distinctive story-telling and poetic expressions of the imagination. The distinctive climates of these northern lands therefore offer physical services and deeper spiritual meanings which are not exposed or valued through routine market exchange mechanisms.

The Inuit have also developed a rich heritage of cultural adaptations to deal with a spectrum of environmental and social change. Local environmental knowledge is an important resource for guiding adaptation to climate change, as evidenced recently, for example, by changed mobility patterns to reduce the risks caused by thinning sea-ice. These Inuit cultures have an ability to learn from conditions of disturbance, to adapt their flexible institutional structures, and to reinvent values in a changing world. The net 'economic costs' of such changes are impossible to gauge.

A different example of how climate and cultural values are deeply entwined, and yet excluded from economic analysis, comes from the Ihanzu people of north-central Tanzania. Rain-making rites and beliefs have been an integral part of many traditional cultures in Africa for a very long time. For the Ihanzu, for example, rain equates to fertility and life; without rain, people, plants and animals cannot survive. Rain-making, or its opposite rain-breaking, is therefore a central feature of the cultural and economic life of these communities. Their rites and rituals are conditioned by indigenous knowledge, tuned over many generations, of the behaviour of seasonal rainfall in this semi-arid part of Africa and also by the social conventions of gender and sexual reproduction envisaged by these people.

Climate, and the continued performance of a 'familiar and understood' climate, therefore has deep symbolic and economic value to the Ihanzu. Changes in the distribution or reliability of rainfall in East Africa would seem to threaten or depreciate such indigenous values. And yet because the relationship of the Ihanzu to their climate is so heavily steeped in constructed conventions, there always remains the possibility for reinvention and adaptation to changed circumstances. Exposure to wider national, or even global, cultural norms and practices forces change on communities such as these, as do changes in economic dependencies. Changes in climate are embedded in these broader altered circumstances, in which ritual, convention and spiritual meaning are continuously changing shape. Representing these deep aesthetic or spiritual values offered by climate, and their change over time, remains beyond the reach of economics.

Economic analyses of climate change are saturated with value judgements and cannot escape the language of ethics. This is not to say that one can say anything one likes about climate change and its economic implications. As with science more generally, the rigour of applying formal concepts and analyses to the questions posed by climate change – in this case economic concepts and analyses – can yield insights about alternative courses of action. It can also expose the places where ethical judgements have to be made. Yet, as economist and international relations expert Michael Toman explains, improved scientific and economic understanding about climate change can 'mask deeper and more complex disagreements about social values. Neither science in general nor economics in particular can resolve the fundamentally moral issues posed by climate change.'[22]

We therefore have to recognise that other approaches to suggesting, negotiating and undertaking action on climate change are necessary – appeals to the scientific authority of the IPCC or appeals to the economic arguments of the Stern Review are, in themselves, far from an adequate basis for agreeing individual, communal, national or global actions.

4.5 Summary

So why do we disagree about climate change? We started this chapter by demonstrating how different economic analyses of climate change, or different interpretations of such analyses, have been used in support of widely different policy prescriptions. Disagreements include support of the Kyoto Protocol or opposition to it; aggressive policy interventions to reduce greenhouse gas emissions within a decade or a gradual and relatively relaxed attitude to climate change mitigation; using cost–benefit analysis to guide policy or rejecting cost–benefit analysis as missing the essence of what climate change signifies.

There is a general recognition from many types of economic analyses that the costs to the global economy of reducing carbon dioxide emissions to achieve concentrations of 550 ppm, or even 450 ppm, are relatively modest – the equivalent of one year's worth of growth in the

---

[22] p. 366 in Toman, M. A. (2006) Values in the economics of climate change. *Environmental Values* 15, 365–79.

global economy, or perhaps two years at the very most. Yet how those reductions are best achieved, and who the economic and political winners and losers might be, is far from clear. And when it comes to determining the damage costs of climate change, economic analyses yield answers that are certainly one, sometimes even two, orders of magnitude apart. We therefore do not agree about the fundamental reasons to take action, nor about the urgency with which we should take action. Is it because we want to maximise our economic well-being, is it because we feel anxious about a planetary climate catastrophe and want to take out an insurance policy, or does it stem from some deeper moral instinct that we are diminishing the quality of life of future generations if we do nothing to slow the rate of climate change now?

We have discussed in this chapter a number of the most important reasons why the economics of climate change, and how it relates to public policy, remains contested. We disagree about climate change because we view our responsibilities to future generations differently, because we value humans and Nature in different ways, and because we have different attitudes to climate risks. More economic analysis with new sets of numbers is unlikely to resolve these differences. In essence, we disagree about climate change because we have different belief systems and because the values we profess seem often different to the values we act on. In Chapter 6: *The Things We Fear*, we will consider in greater depth the question of risk and how we relate to it, but first – in the next chapter – we will explore questions about our belief systems: where they come from, how they differ and whether they might yet converge.

Before we do that, let us make one final observation about economics, climate change and the future. There is a paradox at the heart of all economic analyses about climate change. It is embedded in the Stern Review and in the economic scenarios used by the IPCC. It is the presumption of continued economic growth as measured by conventional GDP. The Stern Review assumes a sustained annual economic growth rate of 1.3 per cent per annum, the IPCC scenarios envisage annual growth rates of between 2.3 and 3.6 per cent. Depending on which scenario is adopted therefore, by 2050 the world economy will be between 70 and 600 per cent larger than it is today and individuals will be between nearly two and six times richer in real terms. Is such an optimistic pro-growth view of the future commensurate with the narratives we hear about depleting natural capital and about the climate risks that future generations may face? What is the purpose of such perpetual economic growth? And why do all analyses of the future make such growth a cardinal assumption?

These are questions that are asked by people of religious faith and by people with none. As ecological economist Clive Spash puts it: 'Traditional pro-growth policies fail to address the problems humanity faces, the necessary transition or the nature of widespread environmental change we are undertaking. All these realisations raise the question of economic activity "for what?" '[23]

Having considered what science and economics can tell us about climate change, and why their powerful yet fractured stories do not lead to an agreed course of action, we will investigate in Chapter 5: *The Things We Believe* what religious and other belief systems can tell us. While they may frequently point to wider questions about the human spirit and the meaning of human life, offering an antidote to the perpetual growth in material consumption criticised above, we will discover that they do not necessarily make it any easier to agree about what to do about climate change.

FURTHER READING FOR CHAPTER 4

Porritt, J. (2005) **Capitalism: as if the world matters**. Earthscan: London.
Jonathan Porritt makes the case for a new turn in global capitalism, a capitalism which places the principles of sustainable development at its centre. He offers a new economics through which he believes prosperity, equity and ecological integrity can be delivered.

---

[23] Spash, The economics of climate change impacts à la Stern, pp. 712–13.

Spash, C.L. (2005) **Greenhouse economics: values and ethics.** Routledge: London.
This book offers a reasonably detailed, yet accessible, account of the basic principles of economics as they apply to climate change. Written by an ecological economist, the book covers the basics of how economics and ethics are inescapably entwined in climate change debates, but does so in a fairly open-minded way.

Stern Review (2006) **The economics of climate change.** Cambridge University Press.
This review of the economics of climate change originated as a commissioned report to the UK Government's Treasury, but is perhaps the most comprehensive such review to date. No discussion about climate change and economics can afford to ignore it. The book reviews some basic principles of economics, illustrates how they can be applied to climate change questions, and undertakes its own quantitative analysis using an integrated assessment model.

Toman, M.A. (2005) Climate change mitigation: passing through the eye of the needle? In: Sinnott-Armstrong, W. and Howarth, R.B. (eds), **Perspectives on climate change: science, economics, politics and ethics.** Elsevier: Amsterdam, pp. 75–98.
This book chapter in an edited interdisciplinary volume about climate change offers a succinct summary of some of the dilemmas faced in undertaking an economic analysis of climate change. It argues for a pluralistic approach to climate change decision making in which ethics and justice become integral to the use of economic tools and frameworks.

FIVE

The Things We Believe

5.1 Introduction

In September 2007 an unusual symposium took place in the icefields of the Ilulissat Icefjord in Greenland. A group of about 200 priests, scientists, theologians and government officials met for a week in western Greenland under the auspices of the organisation Religion, Science and the Environment (RSE); a charitable foundation of the Eastern Orthodox Church. They were there to consider the changing environment in the Arctic and the ways in which humanity should respond. Religious leaders from the Christian, Jewish and Islamic faiths were represented, and the Pope delivered via video a personal message from the Vatican. The venue for the symposium – the outlet of the largest glacier in Greenland – was chosen for its powerful symbolism of the physical changes in climate and the risks associated with a rising sea level.

This is one example of a growing number of initiatives in which leaders of one or more of the world's established religious traditions are forming alliances – either across religious divides or between religious and scientific or secular associations – to call for action to halt the damage that human activities are doing to the natural world, including its climate. These statements, while acknowledging differences of approach and motive, often emphasise that 'we are at one in our belief' that humans must find a way to live more lightly on the Earth, and agree about the profound moral imperative to protect life on Earth. For example, Cardinal Thomas McCarrick, the Pope's representative at the Greenland icefields meeting, led silent prayers for the planet, appealing to everyone to address the impact that humanity is having on the world's climate. 'I don't think we have any differences on this … every religion realises that this world is a gift from God and we have to preserve it.' [1]

But is this true? Are there no differences among the world's religious traditions about climate change, about what it signifies, and what should be done about it? And, even more significantly, are there no differences between the beliefs and spiritual insights of the many citizens of the world who would affiliate with no institutionalised religious tradition? It is only a short step from considering economics – what things we value and why we value them – to examining the role of ethics, spirituality and theology in debates about climate change. And the role of faith communities in this examination is central, as is recognised by many commentators on climate change, including politicians such as the UK's Foreign Secretary David Miliband. 'Climate change is not just an environmental or economic issue, it is a moral and ethical one. It is not just an issue for politicians or businesses, it is an issue for the world's faith communities.' [2]

Our beliefs have a profound influence on our attitudes, on our behaviour and on our politics. Our beliefs determine the sort of world we envision in the future, both the world we would like to inhabit and the world we think most likely we *will* inhabit. While, at some level,

---

[1] Thomas McCarrick, quoted Sunday 9 September 2007. See http://globalwarminglife.com/index2.php?option=com_content&do_pdf=1&id=282 [accessed 10 January 2008].

[2] David Miliband, then the UK Government's Environment Secretary, speaking at the Vatican, May 2007.

Cardinal McCarrick is correct – a basic respect for human life and the created world lies at the heart of all religions – this by no means translates into a common agenda for action. Even if there is some convergence about the moral imperative to 'care for climate', alliances may easily rupture with regard to the apportionment of responsibility for climate change and, even more so, with respect to the urgency, justness and morality of the range of possible responses to climate change. **One of the reasons we disagree about climate change is because we believe different things about our duty to others, to Nature and to our deities.**

This chapter explores the different ways in which the major world religions have engaged with climate change. We also consider how other large-scale collective movements which recognise spiritual or non-material dimensions of reality have done so, for example deep ecologists, eco-socialists, religious humanists. Our beliefs about human dignity and accountability, about the value of the non-human world and about humanity's ultimate purpose and destiny will often shape the way we frame our understanding of climate change. These foundational beliefs drive the category of solutions we bring to the debate and to the negotiating table. Yet these ethical dimensions of the human experience reveal deep social, cultural and religious divides within our world; divides that may often animate other, more visible yet more superficial, reasons why we disagree about climate change.

This is not simply an exercise in exploring the climate change positions of the world's major religions – Buddhism, Christianity (Protestant, Catholic, Orthodox flavours), Hinduism, Islam, Judaism and so on. This will be part of the story, and it will be enlightening. Different faiths, and different traditions within each faith, have engaged at different speeds with climate change, the leaders and laggards revealing much about the institutional power and influence retained by these religions in different national and cultural settings. An interesting exercise in such comparative climate theology was undertaken in 2006 by the Climate Institute in Australia under the leadership of its chief executive Corin Millais. This resulted in a pamphlet being published – *Common Belief: Australia's faith communities on climate change*[3] – in which fifteen different religions summarised their positions on climate change.

But this chapter will also explore the spiritual, moral and ethical dimensions of responses made to climate change in less institutional settings, particularly in the context of traditional cultures, secular justice movements and new environmental theologies. These dimensions are examined in four different areas. In Section 5.2 we first explore why we should care about climate change or, as the American environmental philosopher Dale Jamieson asks: 'What is wrong with climate change?' Quite different theological and ethical arguments can be deployed here, ranging from the deontological (our sense of innate moral duties and rights) to the purely pragmatic or utilitarian (what seems to work best at meeting our needs). To what extent are different religious or secular traditions human-centred, as opposed to planetary-centred or cosmos-centred? And some interpretations of one strand of theology – pre-millennialism in the Christian tradition – suggest that we need not care at all about the fate of the planet or its climate.

We next consider how our beliefs map onto the different ways in which responsibility for climate change can be apportioned (Section 5.3). Even if we all agree that human emissions of greenhouse gases are changing the climate and that we have a duty of care for creation, we emphasise different moral entities when apportioning blame. Some of us blame ourselves for our climate ills. Others blame the rich, the morally degenerate, the capitalist system or our materialist society. We also find that belief systems can profoundly affect the categories of solutions that are advocated for responding to climate change; a relationship examined in Section 5.4. Appeals to the market, to natural justice and to social transformation as solutions may

---

[3] Millais, C. (ed.) (2006) *Common belief: Australia's faith communities on climate change.* The Climate Institute: Sydney.

all find receptive audiences, depending on the particular sets of beliefs held and advocated.

One specific category of solution is developed further in Section 5.5, one which draws its strength from a sense of the counter-cultural and which is rooted in a spiritual ethic rather than in a material one. This is the movement towards simpler lifestyles and thus a rejection of many of the basic tenets of dominant materialist and consumerist ways of living. This has found expression in a number of different faith or spiritual traditions, including, for example, the Amish (as an extreme case) and the deep ecology movement. The argument of this chapter is summarised in Section 5.6. A focus on religious belief and the search for a universal ethic to shape and guide an agreed global response to climate change, important though such a perspective may be, is unlikely to force agreement about what we should do about climate change.

5.2 What is Wrong with Climate Change?

Why are we troubled about climate change? Why do we care about it? Most public opinion polls now suggest that a majority of citizens in many countries around the world want action to be taken to reduce the rate of climate change. People seem to care about prospective changes in climate brought about by collective human actions. But why? What is wrong with climate change?

At a pragmatic level, individuals may worry about climate change if they perceive themselves to be personally disadvantaged in some way. They may sense themselves at risk from increased climate-related disasters, or in some other way see their health, wealth or lifestyle threatened by physical changes in climate. Scaled up to the nation-state, governments may care about climate change because they judge the associated risks are too large to manage safely or because the aggregate economic impact of climate change would threaten their policy commitment to sustained economic growth. This was, in essence, the argument used in the Stern Review that we looked at in Chapter 4: *The Endowment of Value*. We will investigate more of the psychological dimensions of fear and worry in Chapter 6: *The Things We Fear*.

But climate change may also be seen as 'wrong' because of a wider set of ethical considerations: it may contravene our constructed framework of human rights. Climate change may violate – especially for those people who are already easily excluded from society – the basic rights of life, liberty, security, mobility and progeny declared universal by the United Nations in 1948. This claim to care about climate change because of its innate injustice can operate at the level of the individual or at the collective level of the nation-state. Thus a climate risk exacerbated by global warming might cause the death of an individual – the ultimate violation of their rights – and yet, more insidiously, climate change might also lead to the demise of entire nation-states. This latter prospect has certainly been the argument used by the Association of Oceanic and Small Island States when they have brought forward their claims of climate injustice to the fore-front of the world's climate negotiating sessions.

But these examples cited – personal danger, economic loss, and violation of human rights or national sovereignty – do not exhaust the possible reasons deployed by people who view climate change as 'wrong'. We have to dig deeper into the origins of the values that people hold, into the reasons why some things matter and some things don't. And it is here that religious traditions and other forms of spirituality become particularly important. If it is true that climate change is an ethical issue – and few would deny this – we have to engage with the sources of morality and spirituality. We have to listen to what these religious traditions are saying about climate change and we have to understand whether and why people are responding. If Dale Jamieson and other ethicists are right when they claim that there is a fundamental moral intuition at work here – 'that it is simply wrong for humans to causally affect natural systems in such a profound way[4] – then we need to examine from where this intuition originates and how universal it is for humankind. Does it animate only those who believe in the existence of some transcendent deity or reality, or is it also evident in the ethical judgements of materialists and secular humanists?

Religious traditions

All of the world's institutionalised faiths are strong on the duty of care for the created world. There is a reverence for life – a sacredness – that is central to nearly all religious writings, even if expressed in different ways. There is also a belief in the innate value of the entire created order, the material universe brought into being as an expression of the creative will of God, or the gods. These twin tenets of faith are commonly reflected in the idea of 'stewardship', the idea of managing or administering the affairs or property of another person. Humans as viceroys[5] or governors are responsible beings and their responsibility, their accountability, bows ultimately to the divine. The principle of stewardship is common, for example, to Islamic, Christian, Hindu, Sikh, Jewish and Buddhist traditions, and is particularly well expressed by the Quaker tradition: 'We do not own the world, and its riches are not ours to dispose of at will. Show a loving consideration for all creatures, and seek to maintain the beauty and variety of the world. Work to ensure that our increasing power over nature is used responsibly, with reverence for life. Rejoice in the splendour of God's continuing creation.'[6]

---

[4] This position is what is known as 'deontological' ethics. First used in 1930, the term refers to ethical theories that are based on the idea that an action's being right or wrong is intrinsic or intuitive to human reason. It follows that whether a resulting situation is good or bad depends on whether the action that brought it about was intrinsically right or wrong. It contrasts with a 'consequentialist' ethic, where the rightness or wrongness of an action depends entirely on the consequences.

[5] This is the expression as translated from the Qur'an of Islam … 'It is He who has appointed you viceroys of the earth' (Qur'an – 6:165).

[6] *Advices and queries* No. 42 (1994), The Religious Society of Friends (Quakers) in Britain.

This duty of care for the created order, undertaken under the watchful eye of the divine maker, extends most powerfully to the care of other human beings. There is a sanctity to all human life, a call for respect, dignity and value to be the guiding principles by which each person relates to each other. The Dalai Lama expresses it in the language of the Buddhist tradition: 'We are destined to share this planet together and as the world grows smaller, we need each other more than in the past. But ... it is difficult to achieve a spirit of genuine co-operation as long as people remain indifferent to the feelings and happiness of others. What is required is a kind heart and a sense of community, which I call universal responsibility.' [7]

When applied to climate change, these foundational beliefs have frequently been invoked by religious leaders to inspire a larger, more collective vision of our place in the world and of our responsibility to safeguard it. Thus we hear the Catholic Church in the USA claiming that 'global climate change is by its very nature part of the planetary commons; the Earth's atmosphere encompasses all people, creatures and habitats';[8] a claim resonating with Pope Benedict XVI's call for a 'decisive "yes" to care for creation and a strong commitment to reverse those trends that risk making the situation of decay irreversible'.[9] The moral dimension of climate change is argued by many religious leaders to extend beyond our responsibility to people and creatures alive today, to encompass those who are yet to be; as the Archbishop of Canterbury, Dr Rowan Williams, says: 'We are involved in a manifestly unjust situation where those who happen to be alive at the moment are draining off the resources from the vast future community who need to live in a habitable and a just world.' [10]

[7] Dalai Lama from an address 'Seeking the true meaning of peace', San Jose, Costa Rica, June 1989.

[8] Quoted from US Catholic Conference (2001) *Global climate change: a plea for dialogue, prudence and the common good.* Washington DC.

[9] Speech by Pope Benedict XVI, Loreto, Italy, 3 September 2007 – 'Save Creation Day'.

[10] Speech by Dr Rowan Williams, The Archbishop of Canterbury, London, 4 May 2006 – Launch of Tyndall Centre Phase 2.

Secular and traditional beliefs

If these are some of the religious beliefs that may be deployed in calls to resist climate change, or beliefs in which our 'intuitive outrage' against climate change may be rooted, are there equivalent secular beliefs which might similarly inspire and offer the prospect of agreement about how we react to climate change? The most likely place to look for these outside the established religions is in the environmental movement and in traditional cultures.

A number of writers have explored the ways in which religion, science and Nature have met and mingled to give shape to an environmental consciousness. Foremost among these has been the environmental historian Thomas Dunlap. In his book *Faith in Nature: Environmentalism as a religious quest*,[11] Dunlap argues that there are powerful benefits for the future of the world from recognising environmentalism as an emerging religious tradition. While many might react against such a label, Dunlap demonstrates that, at root, environmentalism mimics much that is found in established religious traditions. Three of the most important principles he elaborates are a focus on relationships between humans and Nature; the recognition of the spiritual or transcendent dimensions of the human experience of Nature; and a holistic world-view that, while embracing the power of science, is never able to concede that science can give us access to all that matters.

James Lovelock has similar sentiments when he calls for science to embrace the religious idea of the 'ineffable', of something not to be spoken of because of its sacredness and indescribability. In Lovelock's secular religion, trust is to be placed in Gaia, and ethics are to be rooted in the natural world, which 'has value for and of itself'.[12]

---

[11] Dunlap, T. R. (2004) *Faith in nature: environmentalism as religious quest.* University of Washington Press: Seattle, WA.

[12] pp. 190–1 in Lovelock, J. (2006) *The Revenge of Gaia.* Penguin Books: London.

He calls for a new 'holy book' which sets constraints on human behaviour in the cause of planetary health.

These secular principles of environmental concern, when applied to climate change, suggest the possibility of some bridge-building between the different traditions of religion and environmentalism. If Christians call for a respect for God's world, then environmentalists too recognise the need for respect. If environmentalists recognise the limits of science for determining our right response to climate change, then Buddhists concur that such limits exist and advocate seeking guidance in other places. And as we saw in Chapter 1: *The Social Meanings of Climate*, a romantic ideology of climate reads it as something fragile and precious, something needing to be 'saved' from defilement or destruction. This is language and sentiment which could sit comfortably in many religious traditions, as witnessed below for Catholicism and for Aboriginal beliefs:

> Our very contact with Nature has a deep restorative power – contemplation of its magnificence imparts peace and serenity ... the relationship between a good aesthetic education and the maintenance of a healthy environment cannot be overlooked.[13]

> When the balance that exists between the earth and the people charged with its care and protection is disrupted ... there will be consequences for our society that will inevitably lead to our demise ... we must sustain and nurture the gifts of creation.[14]

Disagreements

So does this brief survey suggest that Cardinal McCormick was right when he proclaimed in Greenland that we have no differences in our beliefs and that 'every religion realises that this world is a gift from God and we have to preserve it'? If the IPCC offers society its most convincing consensus from the scientific reading of climate change,

---

[13] p. 6 in Pope John Paul II (1990) 'New Year message: The ecological crisis: a common responsibility', 1 January 1990, Rome.
[14] Mallais, *Common belief*, p. 6.

do the world's religions offer us a comparable consensus from an ethical or spiritual perspective? Or, less ambitiously, could they? Although it is easy to mine the sacred writings and interpretations of the great religions for common principles, it is unfortunately far harder to synthesize these into a common basis for action, and harder still for individuals and societies to live out any common principles of care. A few examples will be sufficient to show the range of possibilities for disagreement.

In 1967, American historian Lynn White wrote a short article in the journal *Science* in which he argued that monotheist religions, especially Christianity, had been responsible for fostering an anti-Nature ideology which had infused the Western world's attitude to the environment during the Industrial Revolution and beyond. His argument revolved around the notion of 'dominion' found in the Biblical book of Genesis and the mandate for an aggressive stance towards the natural world that this might have appeared to offer humanity. White's thesis has been both supported and attacked by theologians, environmentalists and historians in subsequent decades. It continues to provide powerful ammunition for different positions of advocacy with respect to the role of religion in environmental management, and makes many wary of engaging with certain religious traditions. Environmental decline in Asia has drawn similar concern and comment about the role of oriental religions in shaping certain attitudes towards Nature.[15]

Not entirely unconnected to the above dispute is the value and status to be given to the non-human world in different religious traditions. As we saw in Chapter 4: *The Endowment of Value*, how we place value on entities such as animal species or entire ecosystems cannot be avoided in any discussion about the economics of climate change. One only has to contrast the religious significance of a cow for Hinduism compared to, say, Christianity to appreciate the importance of this

---

[15] See, for example: Dwivedi, O. P. (2001) Classical India, in Jamieson, D. (ed.), *A companion to environmental philosophy*. Blackwell Publishers: Oxford, pp. 37–51.

question. A third example of where disagreement may flourish is a more doctrinal one and concerns eschatology; the system of doctrines concerning the 'final matters' of Death, Judgement and the End of the World. Attitudes to climate change can be significantly influenced by whether, in the Christian tradition, one is an amillennialist or a pre-millennialist (see Box 5.1).

**Box 5.1: American Evangelical Christians
and Climate Care**

Evangelical Christians are a powerful social and political force in the USA. The leading Christian association in the country – the National Association of Evangelicals – claims a membership of 30 million, which represents between 10 and 15 per cent of all eligible American voters.

In February 2006, a new initiative was launched from within the Association aimed at bringing evangelical Christian principles to bear on questions surrounding climate change. Called the Evangelical Climate Initiative, the original eighty-six signatories to the Declaration self-consciously confessed their traditional reticence about engaging with environmental concerns, yet stated that they now saw climate change as a moral issue which American evangelicals had a specific responsibility to attend to. 'Human activity is increasing greenhouse gases in the atmosphere, and the impacts on God's creation and his people will be tragic. To ignore this is unthinkable. To grasp the problem with faith and courage, and with the wind of American ingenuity and goodness at our backs, is morally right and, in our view, faithful to our Creator God.' [16]

---

[16] The Evangelical Climate Initiative, October 2007 press release. See www.evangelicalclimateinitiative.org/pub/ECIMoralLeadership%20Roll%20Call.pdf [accessed 9 July 2008].

The Initiative has had considerable impact in the USA. It has lobbied Washington to take on new climate policy initiatives, it has been welcomed by scientific and business associations who similarly wish to see a stronger US position on climate change, and it has opened up the possibility of inter-faith dialogues about climate change with other religious traditions.

Such a movement demonstrates the power of certain institutionalised religious communities to adopt formal positions on climate change. The Evangelical Climate Initiative even goes so far as to advocate specific principles for climate policy making for the US Administration: an emphasis on free market solutions and the protection of property rights. Yet even with a high profile initiative such as this, and coming from within such a doctrinally driven religious tradition as US evangelicalism, significant disagreements in the movement remain exposed. The Initiative was six years in the making, steering its way delicately through internal theological and political debates, and significant numbers of American evangelical leaders have refused to sign the Declaration. Some of this latter group of church leaders have urged the National Association of Evangelicals to stay out of the global warming debate, stating 'there should be room for Bible-believing evangelicals to disagree about the cause, severity and solutions to the global warming issue.' [17]

Some of this disagreement undoubtedly remains theological. For example, the doctrine of 'dominion over nature' exposed by Lynn White in his 1967 *Science* article (see main text) continues to exert a hold over parts of the evangelical movement. Even more distinctive is the particular form of millennialist theology known as 'premillennialism', the belief that the end of the world and an era of direct divine rule is near. As this era approaches, environmental and social conditions should be expected to decline and political

---

[17] Cited in *Fortune* magazine, New York, 8 February 2006.

turbulence ensue. This belief – as opposed to the less apocalyptic amillennial position – is widely taught across the more conservative end of the evangelical Christian spectrum in the USA. Those who hold pre-millennial views more likely to interpret predictions of climate change – especially catastrophic climate change – as a sign that the millennium approaches. Such presaging of the era of direct divine rule is less likely to stimulate the desire and behaviour to avoid such an outcome.

We next need to enlarge our examination of the role of belief in debates about climate change by considering how these foundational beliefs work themselves out in some of the practicalities and politics of global climate change. We need to show how a convergence of lofty moral principles can easily fragment when challenged by the depth and complexity of the issues raised by climate change. Religious beliefs often do not wield the power needed to bridge differences that exist in our views of science, economics and – as we shall see later in Chapter 9: *The Way We Govern* – politics.

5.3 Theologies of Blame

With ethics being central to all debates about how we respond to climate change, the question of who is responsible for climate change cannot be avoided. Bringing forward for public scrutiny just proposals for responding to climate change requires that we operate under some framework of responsibility and accountability: only from some assumed basis of liability can proposals be evaluated as just or fair. For example, our answer to the question 'Who is responsible for greenhouse gas emissions?' will influence the types of mitigation actions we advocate. Our answer to the question 'Why are some people more at risk from climate change than others?' will similarly shape the types of adaptation responses we propose.

The way we approach such questions of responsibility is strongly guided by our foundational beliefs. Religious, and some secular, traditions may share a basic duty of care for the 'created' world as we have seen, but this does not necessarily translate into a universally agreed apportionment of blame and responsibility for the *lack* of care we exercise. In relation to climate change there may be an initial barrier to overcome: do we believe that humans can affect the climate? The Canadian geographer Simon Donner has argued that many traditional belief systems, whether monotheistic or polytheistic, have long held that the sky is the 'domain of the gods', in other words that weather and climate are beyond the reach and influence of humans.[18] Weather events, especially extreme ones, remain enshrined as 'acts of God' in insurance policies and in popular discourse. Donner reports a survey of Americans following hurricane Katrina in August 2005 which revealed that 23 per cent believed that the hurricane was a deliberate act of God, as opposed to 39 per cent who attributed it to human emissions of greenhouse gases.

Assuming, however, that our beliefs allow us to accept that humans can directly influence the weather, what frameworks of responsibility may be offered by religions? At the most basic level, theological analyses and spiritual traditions can be distinguished according to whether they favour either individual or structural causes for injustice and moral failure. Although there is clearly a wide spectrum of views within each of these categories of agency and structure, we can use this simple dichotomy to illustrate the key point of the argument made here. Differences between theological readings or spiritual intuitions of the relationships of responsibility between God/gods, society, people and Nature – in other words, differences in our foundational beliefs – open up the possibility for considerable differences to exist in the types of responses to climate change we believe are appropriate or necessary.

---

[18] Donner, S. (2007) Domain of the gods: an editorial essay. *Climatic Change* 85, 231–6.

Individual responsibility

Most established religions in their teachings overtly hold in tension the individual (or personal) and systemic (or structural) causes of moral failure. This failure may be couched using the language of sin, injustice, disharmony or evil, but all share the diagnosis that the world we experience, including our experience of ourselves, does not conform to what intuitively we would like it to be. There is a dissonance between our ideals and our experience; for example, the Buddha described 'the human condition by the evocative but untranslatable word *duka* – it means suffering, or unsatisfactoriness or, maybe, lack; the feeling that things aren't quite right.'[19]

This tension between the individual and the systemic causes of 'sin' is visible within the Christian and Jewish faiths, for example, embedded in the many stories of moral failure and injustice recounted in the Old Testament of the Bible. King David is held personally culpable for his illegal appropriation of his neighbour's vineyard, yet the prophet Amos rails against the systemic corruption and exploitation of the poor exercised by elite Israelite society. We repeatedly find, at different times and in different cultures, that one of these dimensions of responsibility for failure is elevated above the other, either through systematic teaching by religious leaders or simply through personal preference.

In relation to climate change, there are many examples of theological emphasis on the individual's responsibility, either for a general uncaring attitude to the natural world or, more specifically, for their emissions of harmful greenhouse gases. The head of the Catholic Church, Pope John Paul II, in his address at the Vatican on World Peace Day in 1990, stated clearly the moral culpability of each human being in relation to their environmental footprint: 'There is an order in the universe which must be respected, and the human person, endowed

---

[19] Vishvapani, 11 December 2007, BBC Radio 4: Today 'Thought for the day'. See www.bbc.co.uk/programmes/b008fphs [accessed 11 November 2008].

with the capability of choosing freely, has a grave responsibility to preserve this order for the well-being of future generations. I wish to repeat that the ecological crisis is a moral issue.' [20] And from the Orthodox wing of the Christian church, through the words in 1997 of Bartholomew I, the Ecumenical Patriarch of Constantinople, is the more explicit charge of sinful personal behaviour in relation to climate change: 'For humans to cause species to become extinct and to destroy the biological diversity of God's creation … to degrade the integrity of the Earth by causing changes in its climate … those are sins.' [21]

Within other theologies these sins can be identified as the sins of individuals in a very precise sense, rather than just the general sins of society. Thus the Anglican Bishop of London, Richard Chartres, claimed in 2006 that 'making selfish choices such as flying away on holiday or buying a large car are a symptom of sin'.[22]

This language of individual moral culpability for the emissions of carbon dioxide has become widely adopted in recent years by a variety of both secular and religious commentators on climate change. Individuals are held accountable for their actions and the services of carbon offsetting companies are held up to be, in a metaphorical sense, equivalent to the selling of indulgences for the remission of punishment for one's sins. 'Just as in the fifteenth and sixteenth centuries you could sleep with your sister, kill and lie without fear of eternal damnation, today you can … drive and fly without endangering the climate, as long as you give your ducats to one of the companies selling [carbon] indulgences.' [23]

[20] Pope John Paul II, New Year message, p. 6.
[21] Bartholomew I, Speech to the Religion, Science and Environment symposium on 'The Black Sea in crisis', 20–8 September 2007.
[22] Quoted in the Sunday Times, 23 July 2006.
[23] p. 210 in Monbiot, G. (2006) Heat: how to stop the planet burning. Allen Lane: London. Monbiot's theology here is not quite accurate, since indulgences in the Catholic tradition never forgave sin, they merely offered remission from the earthly consequences of sins already committed and repented of. Forgiveness remained God's prerogative.

Systemic responsibility

This focus on the individual's responsibility for climate change receives many challenges, however, from both religious and secular perspectives. At the very least, it is claimed, all individuals should not be held equally morally culpable for the emissions for which they are responsible. This was the argument of Indian social scientists Anil Agarwal and Sunita Narain in their powerful pamphlet *Global Warming in an Unequal World* published in 1991.[24] They drew the distinction between 'luxury emissions' and 'survival emissions' in order to argue that one unit of carbon dioxide emitted by an Indian peasant farmer, essential for subsistence, carried a different moral weight to a unit of carbon dioxide emitted by an American tourist flying to the Bahamas. The level of blame is massively different in the two cases. This line of moral reasoning leads quickly into considerations of the economic, social and political structures within which individuals live their lives. Many people feel helpless to make genuinely free choices about their lifestyles and (lack of) consumption options. In one sense the American tourist may be just as trapped in the web of consumption as the Indian farmer is trapped in the spiral of poverty, both caught up in a larger system in which the forces of society condition the individual's consumption.

Most religious traditions are equally well versed in these systemic or structural injustices, which can lead to social evils and which can condition individual moral failure. There is again no shortage of voices that have analysed the causes of climate change in this way and placed blame on the wider political and economic structures of the world. Thus Christian theologian Michael Northcott argues that, far from blaming individuals for the climate crisis, 'Global warming is the Earth's judgement on the global market empire and on the heedless consumption it fosters … accounts of the global warming crisis that

---

[24] Agarwal, A. and Narain, S. (1991) *Global warming in an unequal world*. Centre for Science and the Environment: Delhi, India.

refuse the fundamental conflict between the imperial global economy and the health of the biosphere … cannot do justice to the real roots of the problem.'[25] In a similar vein, although speaking outside any specific religious tradition but with a faith in his own metaphorical construct of Gaia, James Lovelock sees climate change as a result of a systemic failure of humanity to live within the confines of Mother Earth. 'If we fail to take care of the Earth, it surely will take care of itself by making us no longer welcome. Those with faith should look again at our Earthly home and see it as a holy place, part of God's creation, but something that we have desecrated.'[26]

There has been a long tradition of theological critiques of capitalism in its various forms, taking it to task for subverting the principles of divine love, justice and harmony. Liberation theology from within the Catholic Church is one such tradition. Placing the blame for climate change, as Northcott and others do, at the foot of a rampant neo-liberal capitalism is consistent with this tradition. Newer environmental theologies – such as those espoused by post-modern theologian Matthew Fox or Columban missionary Father Sean McDonaugh – extend these insights of the liberationists. It is not just poor and oppressed human beings who are victims of the vicissitudes of climate, but so too is the non-human world, creation itself, a victim of oppressive and destructive human structures and addictions.

Spiritual interpretations of the collective or systemic causes of climate change can also take on other expressions. Although originating in a similar appreciation of the moral dimensions of the relationship between society, Nature and God, these diagnoses of the responsibility for climate change may lead to divergent, if not conflicting, policy responses. They offer the seeds for subsequent disagreement. Thus the Evangelical Climate Initiative in the USA (Box 5.1) steers well

---

[25] p. 7 and p. 269 in Northcott, M.S. (2007) *A moral climate: the ethics of global warming*. Dartman, Longman and Todd: London.
[26] Lovelock, *Revenge of Gaia*, p. 3.

clear of any critique of neo-liberal capitalism and instead concedes, 'Climate change is the latest evidence of our failure to exercise proper stewardship, and constitutes a critical opportunity for us to do better',[27] although it is not clear whether the 'our' and 'us' refer to American evangelicals, Americans in general, or the whole of humanity.

A different angle again is taken by those religious leaders who broaden their analysis and see specific extreme weather events as judgements of God on the immorality and greed of modern society – reflections of the traditional frameworks of moral cause and effect referred to earlier in the essay of Donner. For example, the Right Reverend Graham Dow, the Anglican Bishop of Carlisle in the UK, claimed that the severe flooding in England in June and July 2007 was a 'strong and definite judgement from God … we are reaping the consequences of our moral degeneration … in which every type of lifestyle is now regarded as legitimate … and where economic structures are built on greed. We are in a situation where we are liable for God's judgement, which is intended to call us to repentance.'[28]

5.4 Just Solutions

Our beliefs about the divine, about the spiritual and the transcendent, and about our role in the world as moral agents, shape our sense of duty and responsibility to care for others and for Nature. They affect the way we relate to climate and how we interpret our role in the changes of climate which are occurring. We also hold foundational beliefs about the morality of certain courses of action and about who are the perpetrators of injustices and who are the victims. But we do not all see the world and our relationship with it in the same way. We apportion blame differently. Given that our foundational beliefs do

---

[27] p. 7 in Evangelical Climate Initiative (2006) *Climate change: an evangelical call to action*. Evangelical Climate Initiative: Suwanee, GA.
[28] Quoted in the *Sunday Telegraph*, 2 July 2007.

not always converge, it is not surprising to find that neither do the ways in which we approach climate change nor our responses to it. The categories of interventions, adjustments and solutions that are advocated for managing climate change are often rooted in beliefs which the claims of science, the calculus of economics and the rhetoric of politicians frequently find hard to shift.

To illustrate this point we examine how the complex territory of beliefs and morality explored above maps onto three broad categories of proposed responses to climate change: the correcting of markets, the establishment of justice, and the transformation of society.

Correcting markets

One of the central paradigms for talking about climate change is that of the carbon economy. For the idea of the carbon economy to have traction for policy makers and other social agents, emissions of carbon dioxide and information about emissions of carbon dioxide have to be treated as commodities. Markets for tradable carbon permits and voluntary carbon offsets have to be established. A related idea is that of carbon taxation, a way of internalising into the market economy the externality of climate change. This paradigmatic approach to tackling climate change is the corollary of framing climate change as 'the greatest example of market failure we have seen', as claimed in the Stern Review.

These sweeping ideas for commodifying carbon and globalising its market through either free or regulated trade sit uneasily with many of the beliefs expressed in religious and other spiritual traditions. If the root cause of climate change is seen as a consequence of an overbearing and immoral neo-liberal market economy, further extending the reach and power of that market through carbon trading seems to offer more of the same. If climate change is a consequence of unsustainable material consumption, of 'selfish capitalism' in the phrase of psychologist Oliver James, something deeper is needed than a mere manipulation of the parameters which leaves such consumption intact, even if displaced. As the Western Buddhist commentator Vishvapani puts it: 'Climate change ... will require a fundamental change in society ... the scope of this change challenges the model of ever-increasing wealth on which our society is built. We need help in envisaging a meaningful life and a healthy society that isn't based around consumption. Perhaps that's where religions come in.'[29]

Secular critiques of market-driven consumption can end up at a similar place. A recent report from WWF UK, referring to environmentalism being at a crossroads, suggested that there are inherent contradictions in 'attempting to market less consumptive lifestyles using techniques developed for selling products and services'. An appeal for less consumptive lifestyles is 'unlikely to arise from dependence on a set of material values'.[30] Extending the power of a (modified) market may find favour with certain theological and secular critiques of climate change, but by no means with all.

Establishing justice

A second category of response to climate change is rooted in conceptions of justice, a principled position which might be expected to offer greater resonance with the religious beliefs of various traditions. For many institutionalised and ancient traditions, justice is not a human invention but reflects a divine attribute embedded in the created order. For example, the Hebrew Scriptures call upon Israel to 'seek justice, love mercy and walk humbly with your God'[31] and the followers of the Bahá'í faith are called upon to 'choose thou for thy neighbour that which thou choosest for thyself, if thine eyes be turned towards justice'.[32] Islam has similar impassioned calls for justice: 'O ye who believe! Stand out firmly for justice ... even

---

[29] Vishvapani, 'Thought for the day'.
[30] p. 30 in Crompton, T. (2008) *Weathercocks and signposts: the environment movement at a crossroads*. WWF Report: Godalming, UK.
[31] Old Testament, Micah, chapter 3, verse 6.
[32] Tablets of Baha'u'llah 6.71 and 6.64.

against yourselves, or your parents, or your kin and whether it be against rich or poor.'[33]

Yet operationalising the basic principles of justice in relation to climate change can lead in a number of different directions. Justice would seem to demand that those responsible for altering climate and causing any subsequent damage should be held liable for that damage. This principle underpins part of the framing of the UN Framework Convention on Climate Change, although it leaves much room for debate about the precise apportionment of that liability between present and past generations, and whether it should be individuals, corporations or nation-states which are held liable. Other avenues along which justice might be pursued are those which emphasise procedural or distributive justice – the former making sure that decisions made about climate change are fair and the latter making sure that the actions taken compensate and benefit those who are most disadvantaged by climate change. Climate justice may seem to have much going for it, but whether it is a sufficiently crisp and unambiguous concept around which agreed courses of action will emerge is doubtful. The British social scientist Neil Adger summarises the problem thus: 'It is not surprising that climate justice is contested and means different things for different actors, from parties to the [UN Framework] Convention to national policy constituents and other stakeholders.'[34]

One widely discussed and advocated framework for tackling climate change which claims a strong foundation in the ideas of justice is that of 'Contraction and Convergence' (Box 5.2). Contraction and Convergence (C&C) has been widely endorsed by organisations ranging from the international negotiating bloc, the Africa Group, to the Church of England, and from individuals such as Germany's

---

[33] Qur'an – 5:135.

[34] p. 9 in Adger, W. N., Paavola, J., Huq, S. and Mace, R. J. (2006) *Equity and justice in adaptation to climate change*. MIT Press: Cambridge, MA.

Chancellor, Angela Merkel. Yet despite its alluring logic and appeal to foundational principles of equity and justice, Contraction and Convergence remains marginal to the main international negotiations on climate change.

Box 5.2: Contraction and Convergence:
Who Believes In It?

The idea of Contraction and Convergence (C&C) was first advocated in 1990 by Aubrey Meyer, a UK-based South African musician, author and founder of the Global Commons Institute. It offers a straightforward framework for an international agreement on climate change, the basic principle of which states that all citizens of the world have an equal right in principle to emit greenhouse gases. This right would be granted by a specified future date, the flat individual allowance for each citizen to be derived from an agreed eventual global greenhouse gas concentration target (often claimed by Meyer and others to be around 450 ppm).

The C&C strategy consists of reducing global emissions of greenhouse gases to secure this target level (i.e. 'contraction'); a goal reached because every country brings its emissions per capita to a level which is eventually equal for all countries (i.e. 'convergence'). Initially, these entitlements would reflect the current distribution of emissions across nations, but would eventually converge by a date to be agreed through negotiation (often suggested to be 2050). In order to meet these per capita targets, nations would be entitled to trade allowances, thus facilitating financial transfers from higher-emitting nations to lower-emitting ones. A typical set of carbon emissions profiles under C&C is shown in Figure 5.1.

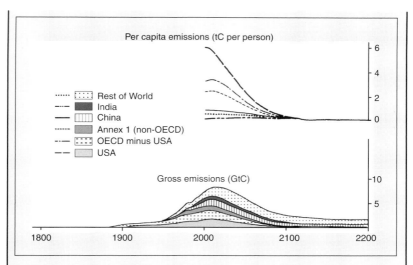

FIGURE 5.1: *Simplified model showing the two key principles of Contraction and Convergence: converging per capita emissions across different world regions (top) and contracting gross global emissions (bottom). Source: Redrawn from the Global Commons Institute.*

There are a number of philosophical and operational difficulties with C&C which have been well rehearsed in recent years.[35] Central to these difficulties is the precise way in which the 'equal allocation' of the emissions right should be defined: Should historical emissions count in the apportionment formula? Should the allowance be emissions per capita or emissions per unit of production? Should land-use emissions be included? Others have criticised C&C for its implicit incentive for further procreation – national emissions

[35] For discussions about the basic principles involved and why they are not self-evident, see pp. 90–91 in Banuri, T. and Weyant, J. (eds) (2001) Setting the stage: climate change and sustainable development. Chapter 1 in *Climate change 2001: mitigation*. Contribution of Working Group III to the Third Assessment Report of the IPCC, Cambridge University Press; and pp. 146–147 in Halsnaes, K. and Shukla, P. (2007) Framing issues. Chapter 2 in *Climate change 2007: mitigation of climate change*. Contribution of Working Group III to the Fourth Assessment Report of the IPCC, Cambridge University Press.

quotas would grow as population expands. This approach also offers little that helps negotiate the two key targets required by the framework: the agreed level of greenhouse gases in the atmosphere and the date by which convergence should occur.

The relevance for the discussion here is that C&C is rooted in a fundamental principle of equality between all human beings, one that should resonate with all religious traditions. It is a human rights-based approach to regulating climate change. For this reason, C&C has garnered significant support from across the religious spectrum, as well as from an assorted variety of other secular and governmental bodies. And yet, despite its claim to rest on 'self-evident' and commonly shared foundational beliefs about justice and equity, the concept of Contraction and Convergence has struggled to be converted into an operational framework either for any national policy or for any international agreement. Neither has it removed the logjam in international negotiations about global climate governance, as we shall see in Chapter 9: *The Way We Govern*.

Transforming society

A third broad category of response to climate change – beyond correcting markets and appealing to principles of justice – is rooted in the idea of social transformation. This emerges from some of the more far-reaching and radical diagnoses of the causes of climate change. Many of these diagnoses are again rooted in religious or other spiritual writings – for example those of Henry Thoreau, Ronald Sider and Bill McKibbin – which share a deep unease with the insatiable materialist consumption of contemporary society. Climate change in this spiritual tradition is not simply a result of market failure, nor a failure to appeal to our instinct for justice. 'At root, the climate crisis isn't just political, economic or even environmental; it's spiritual

… a simpler life that's free of compulsive distraction allows space for awareness, friendship and deeper reflection.'[36]

In its mildest expression, this response to climate change sits comfortably within the reformist elements of the new calls for sustainability – calls initiated most visibly in the Brundtland Commission of 1987 (see discussion in Chapter 8: *The Challenges of Development*). The 'theology' of sustainability has many of its roots in religious movements and in spiritual thinking, for example the writings of philosopher E. F. Schumacher in the 1970s and some of the pronouncements made in the mid-1970s by the World Council of Churches. The latter, for example, claimed that 'a sustainable society which is unjust can hardly be worth sustaining. A just society that is unsustainable is self-defeating. Humanity now has the responsibility to make a deliberate transition to a just and sustainable global society'. [37] The transformation of society implied by these calls goes well beyond putting a price on carbon or introducing explicit principles of justice into international trade and diplomacy. If Northcott's analysis of climate change from a Christian perspective is to be adopted, then something much more far-reaching is called for; the reform of global capitalism using 'kinds of politics … which will need to engage every citizen, household, corporation and organisation'.[38]

This is the sort of radical bottom-up transformative vision that the writer and entrepreneur Paul Hawken speaks of in his account of 'how the largest movement in the world came into being and why no one saw it coming' in his 2007 book *Blessed Unrest*.[39] Hawken claims that the idiosyncratic founders of these new networks of social movements can be traced back metaphorically to the healers, priestesses,

---

[36] Vishvapani, 'Thought for the day'.

[37] World Council of Churches, 1976: quoted in Langhelle, O. (2000) Why ecological modernisation and sustainable development should not be conflated. *Journal of Environmental Policy Planning* 2, 303–22.

[38] Northcott, *A moral climate*, p. 281.

[39] Hawken, P. (2007) *Blessed unrest: how the largest movement in the world came into being and how no-one saw it coming*. Viking: New York.

philosophers, monks and poets of earlier eras. This is bottom-up community power set against top-down privileged power, climate change demanding something which is not in the gift of the market, the politician, the diplomat or the celebrity to offer. We will consider in the next section where one of the well-springs for this type of transformation may be found.

We end this section by noting how fragmented and divergent are the three categories of response to climate change we have examined here – markets, justice, lifestyles – yet each can claim roots in foundational beliefs and religious sentiments.

5.5 Personal Transformation

We saw in Section 5.3 how different lines of religious thought place the blame for climate change variously on the individual's conduct or on the collective structures of the world, whether these be economic, political or social. A belief in the individual's moral responsibility for climate change might suggest that an adequate response to climate change must include, at the least, a transformation of the individual's behaviour. This is a position which also overlaps with those who call for a restructuring of society to diffuse the drivers of climate change. It emphasises the spiritual over the material, the community over the consumer, and the egalitarian over the hierarchical. Since these are areas into which religious and spiritual traditions often speak, we might ask whether this response to climate change offers a shared basis for agreement. Let us examine two aspects of this pathway towards personal transformation: the idea of, or search for, inner contentment or well-being; and the outworking of this search in model communities.

Well-being

Most religious traditions are very clear that the essence of contentment and human well-being is to be found in non-material values rather than through material consumption. Thus the imperative spoken by Jesus Christ to 'seek first the kingdom of God and his righteousness and all these things shall be given unto you[40] and the Islamic idea of *fitra* – humanity's natural, but now lost, state of contentment[41] – both speak of such values.

These long-standing wisdom claims have been subject to empirical social science in recent years and been found to accord well with observable and lived experience. These non-material values are what psychologists refer to as 'intrinsic' as opposed to 'extrinsic' – for example, personal growth, community involvement and relationships rather than wealth, status and power. For example, at a population level, correlation between gross national product (GNP) per capita and life satisfaction can only be found to a certain level of material consumption. Above that level, further growth in GNP per capita has no statistical influence on indicators of subjective well-being; in other words, 'happiness' (Figure 5.2). The support of family, friends and community, a purposeful role in life, and basic religious and political freedoms seem to be much more important determinants of subjective well-being above a certain – quite modest – level of material consumption.

This exposure of the false presumption that conspicuous and increasing material consumption equates to well-being has been used widely by those advocating that both the diagnosis and the solutions to climate change are rooted in choices about lifestyles, human values and spirituality. Spiritual leaders persist in their teaching of such wisdom and many social commentators draw attention to this reality – for example Oliver James in his best-selling 2007 book *Affluenza: How to be successful and stay sane.* Yet these insights into the sufficiency of 'intrinsic' values sit uncomfortably alongside the contradictory 'extrinsic' values demanded (whether implicitly or explicitly) by conventional economics, commercial marketing and political

---

[40] New Testament, The Gospel of Matthew, chapter 6, verse 33.
[41] See Chishti, S. (2003) *Fitra: an Islamic model for humans and the environment*, in Foltz, R., Denny, F. and Bahaaruddin, A. (eds), *Islam and ecology: a bestowed trust.* Harvard University Press: Cambridge, MA, pp. 67–82.

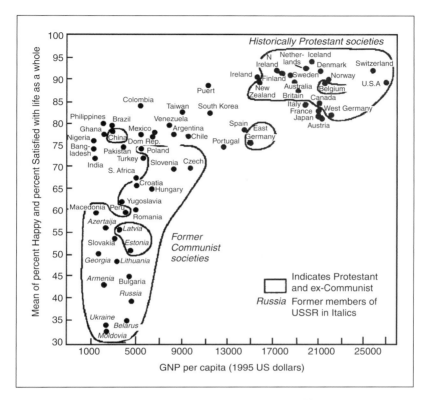

FIGURE 5.2: *Level of economic development as measured by GNP per capita plotted against a measure of subjective well-being for various countries. Two subgroups of countries are identified: historically Protestant cultures and ex-Communist cultures. Data are for the 1990s.*
Source: *Inglehart and Klingemann (2000).*

rhetoric. Such traditional wisdom seems opposed to our expressed values as revealed, for example, through collective human behaviour.

Those in the West (especially) who seek to live low-carbon domestic lifestyles – low-energy light bulbs, locally sourced food, driving hybrid cars – and yet who holiday two or three times a year at destinations at the end of an international or intercontinental flight, reveal the psychological tension which this clash of values can lead to. Is minimising climate change a matter of reducing carbon emissions while leaving most other aspects of our lifestyles untouched or, as this wisdom tradition might claim, a question of purging a deeper desire for profligate and unsustainable consumption? Eamon O'Hara, an Irish policy adviser in Brussels, claims the latter and expresses it this way: 'How many people are tired and weary of modern living? The endless cycle of earning and consumption can be exhausting and does not necessarily bring happiness and fulfilment. Can we do things differently and better? If we don't, then we are headed for certain disaster, regardless of whether or not we manage to reduce our emissions.'[42] It is this tension between consumption and contentment which many individuals experience that leads some to advocate more radical communitarian lifestyle solutions to climate change.

Model communities

The ultimate expression of a shared way of life which sits lightly on climate and the physical world and that emanates from a deeper spiritual philosophy can be found in some religious and secular communities. The Christian Mennonite tradition is one of these in which

> if we are ever able to stop destroying our environment, it will be because person by person we decide … to turn aside from greed and materialism. It will be because we learn that joy and fulfilment come through right relationship with God, neighbour and earth, not an ever escalating demand for more and more material consumption. Nowhere is that more possible than in local congregations that combine prayer and action, worship and analysis, deep personal love for the Creator and for the Creator's garden.[43]

The Amish, Hutterites and the Franciscan and Benedictine Orders are other faith communities which have emerged at different times over recent centuries from a broadly Christian tradition and which now

---

[42] O'Hara, E. (2007) Focus on carbon 'missing the point'. BBC News On-line, Viewpoint, 8 August 2007. See http://news.bbc.co.uk/1/hi/sci/tech/6922065.stm [accessed 9 July 2008].

[43] Professor Ronald Sider, Palmer Theological Seminary, Pennsylvania, and Founding President of the United States Evangelicals for Social Action.

practise communal living with low levels of material consumption and small ecological or carbon footprints.[44] Amish values, for example, translate into a simple lifestyle: small-scale agriculture, separation from the world, pacifism, and a strong work ethic; values rooted in earlier sixteenth-century radical Protestantism. Their use of transport, energy and (most importantly) their agriculture reveals a radical sustainability and sympathy with climate, wildly at odds with prevalent Western culture and values.

These examples are cited to argue that commonly shared beliefs which spring from a religious or secular spirituality, when put to work in designing responses to climate change, can lead to deeply divergent prognoses for salvation. Some have argued that the growing engagement of religious leaders and communities with climate change offers opportunities for finding common solutions – the sense that 'there is an overarching spirit of looking past differences, reaching out across faith communities and arising to realise common goals'.[45] I suggest here, however, that the gap between shared principles and outlooks, on the one hand, and advocated policy and manifest behaviour, on the other, offers up as many opportunities for disagreement about climate change as for agreement.

5.6 Summary

So why do we disagree about climate change? Climate change is increasingly discussed using language borrowed from religion, theology and morality. Those in the West hear of 'penances' being paid for 'carbon sins' through the purchase of emissions offsets, read of the

---

[44] Vonk, M. A. (2006) The quest for sustainable lifestyles and quality of life: contributions from Amish, Hutterite, Franciscan and Benedictine philosophy of life. Paper presented at the conference 'Exploring religion, nature and culture', Gainesville, FL, 6–9 April 2006.

[45] p. 39 in Posas, P. J. (2007) Roles of religion and ethics in addressing climate change. Ethics in Science and Environmental Politics 6, 31–49.

personal 'guilt' associated with flying, and are challenged to 'repent' of their profligate consumption and to 'convert' to low-carbon lifestyles. There is even a new missionary movement of communicators and advocates commissioned on the back of Al Gore's evangelical movie *An Inconvenient Truth*.[46] As awareness of climate change penetrates further into social and cultural life around the world, how belief systems and ethics inflect the public discussion of climate change, and our response to it, will become increasingly important.

Far from being a simple problem of science (Chapter 3) or of economics (Chapter 4), there is a deepening appreciation that climate change – both in the way we frame it and the way we define our response – can only be grasped through appreciation of its ethical dimensions. This has been well put by the Anglican Bishop of Thetford, David Atkinson. 'Climate change is … opening up for us … questions about human life and destiny, about our relationship to the planet and to each other, about altruism and selfishness, about the place of a technological mind-set in our attitude to the world, about our values, hopes and goals, and about our obligations for the present and for the future. These are moral and spiritual questions and therefore theological ones.'[47] During the Tenth Meeting of the Conference of the Parties in Buenos Aires in December 2004, some of these ethical dimensions were articulated in the 'Buenos Aires Declaration on the Ethical Dimensions of Climate Change'.

But it is one thing to recognise the inescapable ethical character of climate change debates. Quite another is to find ways of reconciling what can be apparently contradictory ethical stances which themselves may be traced back to divergent interpretations of foundational religious beliefs and spiritual intuitions. As we have seen in earlier

---

[46] Haag, A. (2007) Climate change 2007: Al's army. *Nature* 446, 723–4. The relevant quote here is: 'Gore calls the trainees his "cavalry", but a more apt name might be missionaries, given the fervour with which they approach their roles.'

[47] p. 28 in Atkinson, D. (2008) *Renewing the face of the Earth: a theological and pastoral response to climate change*. Canterbury Press: London.

chapters, appeals to science or economics for moral authority or guidance are usually of limited effect, even if some of these ethically loaded questions are clothed in the language of science (e.g. 'What is dangerous climate change?') or economics (e.g. 'What is the monetary value of a human life?'). What this chapter has shown is that appeals to religion – to arguably, or hopefully, common spiritual and human values – may also be a long way from reconciling a fragmented and argumentative world.

Those who argue that climate change presents us with a unique opportunity to seek out shared ethics in an interconnected world and then deploy these ethics on a global scale to diffuse our fears about climate change are well intentioned. But I suspect that the fissures and canyons that quickly start opening once we descend from the high and lofty peaks of religious sentiments about 'care for creation' and 'respect for life' are not ones that can easily be bridged. Such advocates must find positive answers to the following questions, not just statements of intent, but demonstrations of significant-scale successes in redirecting the momentum of society. Are faith-based coalitions of activists more likely than secular movements to bridge cultural and economic divides between East and West or between North and South? Can evangelical Christians work alongside liberal environmentalists to secure climate change policy goals? Science has been effective at identifying the physical reality of changing climates, but can science ally with faith communities and other ethical movements to provide a platform for enabling effective responses?

Those whose beliefs lead them to propound a strictly materialist view of reality have perhaps even further to go in seeking a mechanistic explanation of how a universal morality for guiding us through a century of changing climate may emerge. Yet the insights of evolutionary biologists and psychologists can be important here. They offer a credible, if incomplete, material explanation of cultural evolution and moral intuition and its social functions. American psychologist Jonathan Haidt has optimistically argued: 'Because morality may be as much a product of cultural evolution as genetic evolution, it can change substantially in a generation or two … as technological advances make us more aware of the fate of people in faraway lands, our concerns expand and we increasingly want peace, decency and co-operation to prevail in other groups and in the human group as well.'[48]

Recent human history is not auspicious in this regard, however, and with respect to offering a pathway towards a new ethic for climate sustainability, Haidt and others like him will have to defend their faith a while longer.

We have examined the offerings of the three great human practices of science, economics and religion with respect to climate change. Having argued that each of these human enterprises contains logical or operational weaknesses which limit their ability to forge agreement between us on what to do about climate change, we shall now start a second round of investigations into the reasons why we disagree about climate change. We will start with the ideas of risk and fear, and consider to what extent we may inadvertently be constructing the problem of climate change to be worse than it need be. The psychologists will be our first companions in this investigation.

FURTHER READING FOR CHAPTER 5

Dunlap, T.R. (2004) **Faith in nature: environmentalism as religious quest**. University of Washington Press: Seattle, WA.
This beautifully written essay explores the origins of the environmental movement in the USA and traces the parallels between its principles, attributes and goals with those of institutionalised religion. Dunlap's insights can be directly read across from environmentalism into debates about climate change, where science, values, beliefs and action similarly become entangled. His argument is that recognising these entanglements is liberating rather than restricting.

---

[48] p. 1001 in Haidt, J. (2007) The new synthesis in moral psychology. *Science* 316, 998–1002.

Millais, C. (ed.) (2006) **Common belief: Australia's faith communities on climate change.** The Climate Institute: Sydney.
A short guide to statements issued by the Australian leaders of fifteen distinct religious faiths or traditions about the nature of climate change and their proposals for personal and social responses. These faiths include Aboriginals, Christians, Buddhists, Hindus, Muslims, Sikhs and Jews. This is a very useful starting point for dialogue and for shared respect between faith communities. But it also shows that a religious synthesis or integration about climate change and developing a high-level shared ethical position on climate change is still awaited.

Northcott, M. S. (2007) **A moral climate: the ethics of global warming.** Dartman, Longman and Todd: London.
Written by an ethicist, theologian and Episcopalian priest, this book examines the basics of climate change science, politics and economics and, in particular, a variety of theological and ethical reactions to the dilemmas posed. Northcott's diagnosis of climate change is blunt and direct. His challenges for corrective actions are equally blunt. These reveal his strong Christian conviction that doctrine and practice are inseparable, which at times lends the book the air of a preacher's polemic.

Taylor, B. (ed.) (2005) **Encyclopedia of religion and nature.** Thoemmes Continuum: London and New York.
This is a massive encyclopedia covering an enormous breadth of thought, scholarship and action concerning the relationship between religion and nature, but the volume is prefaced with an excellent opening twenty-page essay by the editor Bron Taylor. This essay introduces the nature of religion(s), their relationships with Nature and the changing ideas of Nature, and offers a good quick guide to how these have evolved in recent centuries. There are, of course, many other more specialist entries that could usefully be read as well.

SIX

The Things We Fear

6.1 Introduction

> Time is running out, and fast. Rising carbon dioxide levels and higher temperatures will soon set in motion potentially catastrophic changes that will take hundreds or even millions of years to reverse.

This was the way in which the outcome of an international conference on Avoiding Dangerous Climate Change held in Exeter, UK, in February 2005 was reported in the British science magazine *New Scientist*.[1] Under headlines such as 'The edge of the abyss' and 'Act now before it's too late', the science reported at the conference rapidly found its way around the world and into the cabinet rooms of governments, the board rooms of companies, and the living rooms of the public. A month or two later, the US weekly news magazine *Time* ran a front cover special on global warming with an equally intimidating title: 'Climate change: be worried, be very worried'.[2]

The Exeter Conference signalled a step-change in the ways in which the risks associated with climate change were conceived, presented and debated in the public sphere. Previously, climate change

---

[1] *New Scientist*, London, 12 February 2005 and 26 March 2005.
[2] *Time* magazine, New York, 28 April 2005.

had usually been discussed in terms of incremental changes to the average conditions of climate; incremental changes to which it might – at least in some regions and with some foresight – be possible to adapt. There had also been attention paid to changes in the likelihood of the extremes of weather. As average climate conditions changed, the chances of weather extremes such as heatwaves, storms and heavy rainfalls – and their attendant risks for society – might increase in a disproportionate manner. But the Exeter Conference opened up to a wider public a third category of climate change risks: abrupt or rapid changes in the climate, ominously described by *New Scientist* as 'waking the sleeping giants' of the Earth system; giants such as melting polar ice sheets, the collapse of the Atlantic Ocean heat conveyor, and the exhaling of methane burps from the frozen tundra.

This conference of scientists had been called in October 2004 by the UK Prime Minister Tony Blair as part of his Government's preparations for the G8 Summit which he was to host in Gleneagles in July 2005. Blair called upon the world's climate scientists to 'identify what [concentration] of greenhouse gases in the atmosphere was self-evidently too much', an interpretation of Article 2 of the United Nations Framework Convention on Climate Change (UNFCCC) which laid out the objective of avoiding dangerous human interference with the climate system. Identifying such a concentration in an international scientific setting would, presumably, help focus the minds of the G8 leaders on what action to take about climate change. While the scientists at Exeter adopted no formal 'level' of danger, the exchanges at the meeting tended to converge around a concentration of greenhouse gases that would offer a fair chance of limiting global warming to no more than 2°C above the pre-industrial temperature.[3] As

[3] This figure also happened to be the stated policy goal of the European Union (since 1996) and, by association and direct confirmation, the stated goal of national UK climate policy.

John Schellnhuber, one of the scientific convenors of the conference, claimed: 'If we go beyond two degrees we will raise hell.'[4] The risks reported at the conference were further amplified by relativising them to earlier assessments: they were now 'more serious than previously thought'.

As we saw in Chapter 1: *The Social Meanings of Climate*, there is a long history of humans relating to climate in pathological terms. A prospective, and not fully predictable, change in climate offers fertile territory for the heightening of these fears. Recent narratives of climate change seem to be part of this history. Climate risks surprise us and shock us; both the risks that we see or experience today and the risks in the future that scientists tell us we need either to avert or to prepare for. In recent years, the risks associated with impending future climates have been increasingly communicated using the language of disaster, catastrophe and terror. This selection of quotations from politicians, campaigners, scientists and commentators illustrates the tendency:

Climate change over the next twenty years could result in a global catastrophe costing millions of lives in wars and natural disasters.[5]
If you want to be frightened about anything, you want to be frightened about the impact of climate change. It's worrying for our generation – it's even more worrying for the generation coming behind.[6]
The impact of terrorism affects hundreds or thousands or maybe, if they're really good at it, millions. But the product, probability and impact of climate change is greater.[7]

---

[4] *New Scientist*, London, 12 February 2005, p. 12.
[5] The *Observer*, 22 February 2004, summarising the report by Schwartz, P. and Randall, D. (2003) *An abrupt climate change scenario and its implications for US national security*. Report by the Global Business Network prepared for the US Department of Defense.
[6] Alastair Darling, *The Guardian*, Financial Pages, London, 28 June 2006.
[7] Chris Rapley, Lecture at the National Research Council, Ottawa, October 2006.

Terror only kills hundreds or thousands of people. Global warming could kill millions. We should have a war on global warming rather than the war on terror.[8]

This chapter examines the construction of risk and its significance in discourses around climate change, drawing upon ideas from social and behavioural psychology, risk perception and Cultural Theory. At its simplest, the risk of an event can be described as a function of both the probability of the event's occurrence and the magnitude of the event's impact. As we saw in Chapter 3: *The Performance of Science*, in the case of climate change neither of these numerical quantities are uncomplicated to determine. The significance of risk for the formation of public policy is further complicated by a third element in the equation, which is the individual or collective *perception* of risk. No two people necessarily have the same perception of a given risk and certainly do not necessarily evaluate its significance for individual or collective behaviour in the same way. The overall concept of 'risk' is therefore influenced by a series of considerations that take us a long way from the pronouncements of the scientists at Exeter about the probability of the 'sleeping giants' awakening. **One of the reasons we disagree about climate change is because we evaluate risks differently.**

We start the chapter by investigating different ways in which risks can be constructed and the various influences that affect their construction (Section 6.2). We examine the peculiar risk characteristics of climate change and contrast the idea of risk constructed through mental processes with risk constructed through cultural influences. Beyond the scientific elements that inform the description and assessment of climate risks made by the Intergovernmental Panel on Climate Change (IPCC), influences such as personal experience, values and

[8] Stephen Hawking, quoted in *The Times*, London, 31 January 2007.

world-view all play important roles in ascribing significance to any given risk.

At the centre of debates about climate change is the idea of 'dangerous climate change', and Section 6.3 explores what this idea might mean and how it relates to the different ways in which risk can be constructed. Who defines danger? Can dangerous climate change be discovered by science or can it only be encountered through experience? This brings us on to considerations of ourselves and how we think of our own vulnerabilities and our own resilience in the face of climate risks (Section 6.4). We examine differences in the way that individuals, social groups and national publics perceive and evaluate climate risks, which may help go some way to understanding why we understand climate change differently. It may also help to reveal why gaps exist – sometimes very large gaps – between the apparent status of climate change risks and the changes in behaviour that may be needed to mitigate those risks.

Building on the idea of risk as being socially constructed, we finally consider the ways in which climate risks get amplified (or attenuated) within society (Section 6.5). The roles of different interest groups, different types of media, and different political voices all become important in understanding why some risks seem larger than others and why some risks rather than others are acted on by governments. In all of this, the communication of risks, and ultimately the communication of the scientific knowledge that contributes to the perception of risks, is seen to be central. And this leads us into Chapter 7: *The Communication of Risk*.

6.2 Cultures of Risk

'Can we know the risks we face, now or in the future? Are dangers really increasing or are we [just] more afraid?' These two questions were posed by anthropologists Mary Douglas and Aaron Wildavsky at the beginning of their 1982 book *Risk and Culture*,[9] in which they laid out a cultural theory of risk perception. They questioned the presumed scientific objectivity of environmental risk assessment and instead showed that cultural, political and psychological factors play a more dominant role in the way risks are constructed, perceived and ranked by people living in societies. Although originally writing well before climate change had emerged as a major public issue of risk and safety, their insights into how risks are perceived by individuals are crucial for helping us understand why today we worry differently about climate change.

A conventional approach to the idea of risk defines it as the product of the likelihood of an event occurring and its magnitude or impact. In principle, these two elements in the risk equation – likelihood and impact – can be quantified using the tools of science and economics. Thus, for example, the risk of a hurricane might be compared to the risk of a heatwave. Meteorologists are asked to determine the likelihood of events of a certain extremity occurring and economists are asked to determine what the consequential damage for society might be. Multiply these two quantities together and one has an estimate of the 'objective' risk. But we have seen in earlier chapters why this is not so easy to do in the case of climate change. Meteorologists have to grapple with deep uncertainties in their predictions of future climate change, and economists have to struggle with judgements concerning values and ethics. Even in such a simple formulation of risk we encounter serious problems in putting numbers into the equation.

Nevertheless, and as we have already considered, the IPCC has sought to constrain the uncertainties about future climatic events unfolding, while economists – those for example of the Stern Review – have sought to place a damage cost on their occurrence.

9 Douglas, M. and Wildavsky, A. (1982) *Risk and culture: an essay on the selection of technological and environmental dangers.* University of California Press: Berkeley, CA.

Read these two scripts together and it would seem that the risks of climate change are serious, large, and reasonably well identified. Add substantial buy-in from governments around the world, together with the endorsement of the United Nations, and this would surely suggest that a large majority of people on the planet should be fully aware of the risks of a changing climate and be acting accordingly. But is this all there is to understand about risk? Is it really as simple as quantifying the risk, communicating the resulting assessment clearly and widely, and thereby achieving a common and universal appreciation of the dangers that we face?

It is here that a third question posed by Douglas and Wildavsky in 1982 becomes important. 'How do people decide which risks to take [or to mitigate in our case] and which to ignore?' If the idea of risk was as neat as the formula outlined above suggests, then we would all share a common rank ordering of risks faced. Through our behaviour we would then seek to avoid or mitigate the gravest risks, while happily living with those risks which carried little weight for us.

What we would have failed to consider, however, and what Douglas and Wildavsky elaborated in their book, are the wider cultural and psychological contexts which influence the way in which individuals perceive particular risks. Risk assessment, or risk quantification, is not the sole preserve of the expert or the scientist. Such analysts may have specific knowledge and technical skills that the majority of citizens do not have, and they may be able to evaluate different types of risk, but in the end it is individuals who have the last word in any risk assessment. It is in the differential *perceptions* of risk that the scope for ultimate disagreement lies. And it is cultural, social and psychological conditioning that will have the greatest influence on these individual risk perceptions.

Hazards or behaviour which threaten valued assets or lifestyles will be viewed by a society or culture as risky. But since each culture holds different assets and lifestyles to be of value, they will assign varying levels of importance to different risks. Each society, each social group within a society, and even each individual within a social group, may prioritise quite differently which risks to avoid and which risks to live with. For example, whether a summer heatwave is regarded as a personal risk may depend not just on age and physiology, but also on occupation, attitude to leisure, and belief in the dangers of exposure to high levels of ultraviolet radiation.

Some have argued that one of the most likely factors determining environmental risk perceptions, including the risks of climate change, will be economic wealth. The American political scientist Ronald Inglehart did much to promote this view with his idea of 'post-materialist values'.[10] He suggested that concern for the environment was a luxury good, an object 'purchased' only after other, more basic, needs had been met. According to this hypothesis, climatic risks were likely to be viewed with more serious concern in the wealthier nations of the North and, within those nations, to be viewed most seriously by those individuals with the highest material standard of living.

But differences in risk perception are unlikely to be as easily explained as this. Douglas and Wildavsky developed a much broader framework, later elaborated by anthropologist Michael Thompson, for thinking about why people graded risks in the way they did. Their framework, which has become known as the 'cultural theory of risk', goes beyond superficial categories of rich and poor, or politically left-leaning versus right-leaning. Instead, the way risks that are perceived and given priority is a function of how individuals see themselves in relation to others in society, which in turn affects their world-view, their values and their 'ways of living'.

---

[10] See, for example, Inglehart, R. (1997) *Modernisation and post-modernisation: cultural, economic and political change in 43 countries.* Princeton University Press: Princeton, NJ.

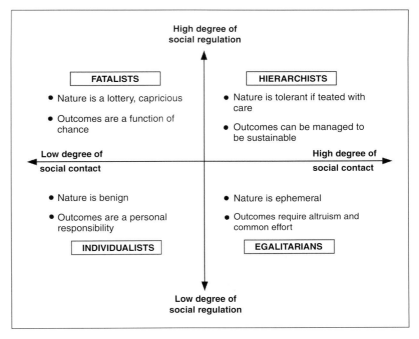

FIGURE 6.1: *The four 'ways of life' derived from Douglas and Wildavsky's original (1982) Cultural Theory.*
Source: adapted from O'Riordan and Jordan (1999).

The framework of Cultural Theory yields a four-fold classification into which individuals, social groups or entire societies may be placed. The classification draws upon two fundamental dimensions: the extent to which people are group-oriented or individual-oriented, and the extent to which people believe that many rules are needed to control behaviour or that only few rules are necessary. In this classification (Figure 6.1), both 'hierarchists' and 'egalitarians' share a sense of solidarity as members of society; they are strongly group-oriented and feel bonded to larger social units. They differ, however, in whether they see this social bonding as primarily vertical (the hierarchist) or horizontal (the egalitarian). The hierarchists see a strong social structure of rank, role and place, with social interaction governed by multiple sets of rules. The egalitarians view everyone as fundamentally equal, joined together through purely voluntary associations with few governing rules.

The other two categories – 'individualists' and 'fatalists' – have a different and less group-oriented view of society. In both cases, they see little or very weak social bonding within society. The individualists additionally see little need for any social structuring of relationships around conventions or rules, whereas the fatalists accept their position as isolated individuals, but within a stratified and rule-bound society.

Douglas and Wildavsky argue that these four 'ways of life' – these different views of the relationship between the individual and society – offer a much richer insight into how and why different people in different cultures perceive risks in different ways. By bringing cultural and psychological insights to bear on the question of risk perception, they take the argument away from the objective analysis of risk, or indeed from Inglehart's purely economic argument that material wealth can explain the diversity of risk perceptions that exist. And as Box 6.1 shows, the Cultural Theory of risk can be mapped onto different ways of regarding Nature – i.e. Nature as either benign, ephemeral, perverse/tolerant or capricious. These insights can, in turn, be used to explore the ways in which the risks associated with climate change are perceived and talked about. They may also offer some help in understanding why different types of responses to climate change may be favoured over others. For example, the idea of the precautionary principle that we introduced in Chapter 4: *The Endowment of Value*, as a framework for thinking about climate policies is likely to find greater resonance among egalitarians, with their view of Nature as being ephemeral or vulnerable, than among individualists, who regard Nature as benign or resilient.

Box 6.1: The Myths of Nature and the Perception of Risks

The idea that the position of people in Douglas and Wildavsky's map of cultural space influences the way in which they perceive risks associated with climate change finds support in other areas of study. We explored in Chapter 1: *The Social Meanings of Climate*, the different ways in which societies in the past and individuals in the present have conceived of the idea of climate, and how these conceptions may be influenced by certain views of Nature. In the 1980s, ecologist Buzz Holling sought to classify different implicit views of Nature held by Canadian forest managers and to relate them to wider conceptions of whether the Earth is seen as stable or unstable. He called these the 'myths of nature'. His three myths, with a fourth added later by Michael Thompson, can be mapped onto the four 'preferred ways of life' first proposed by Douglas and Wildavsky in 1982. These myths, or world-views, also help us to navigate our way through the different responses people have to the idea of climate change, to the risks it presents to humanity, and to the means and urgency by which climate change may be managed.

Holling used the iconic device of depicting his nature myths using the picture of a ball in a landscape. These icons are reproduced in Figure 6.2, along with brief descriptions suggesting how these different world-views and nature myths map onto different perceptions of the risks of climate change.[11] The illustrative

[11] These descriptions draw in part upon Thompson, M. and Rayner, S. (1998) Cultural discourses, in Rayner, S. and Malone, E. L. (eds), *Human choice and climate change. Vol. 1: The societal framework*. Battelle Press: Columbus, OH, pp. 265–344.

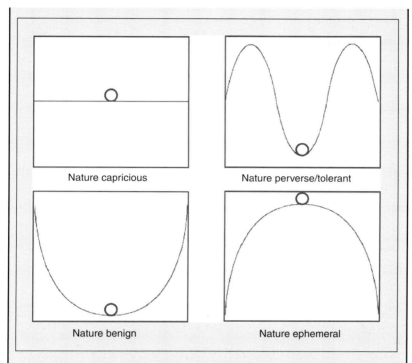

Nature capricious

Nature perverse/tolerant

Nature benign

Nature ephemeral

FIGURE 6.2: *Schematic representations of the four myths of Nature, originally suggested by Buzz Holling.*
Source: adapted from Holling (1986).

stereotypical quotations in each case are extracted from the analysis of the linguistic repertoires of climate change found in the UK print media between 2005 and 2007.[12]

Nature as benign. This myth is favoured by those who fall into the 'individualist' category. It tends to view the climate system as favourably inclined towards humanity, at least within broad

[12] Ereaut, G. and Segnit, N. (2006) *Warm words: how are we telling the climate story and can we tell it better?* IPPR: London; and Segnit, N. and Ereaut, G. (2007) *Warm words II: how the climate story is evolving.* IPPR/Energy Savings Trust: London.

definable limits. Any risks introduced by climate change are viewed as manageable and, even with humans altering the global atmosphere, the Earth's climate will re-establish itself at a tolerable and non-dangerous level. Such an individualist might comment: 'We will adapt and change by embracing the [climatic] future rather than trying to turn back the clock of technological change.'

Nature as ephemeral. This myth is favoured by those who fall into the 'egalitarian' category. In direct contrast to seeing Nature as benign, this view of the climate system sees it as existing in a precarious and delicate state of balance. The slightest perturbance by humanity can trigger a collapse in the system. The risks of climate change are frightening and may spiral out of control. Such an egalitarian might comment: 'The collapse of civilisation is the outcome if fossil fuel use rises.'

Nature as perverse/tolerant. This myth is favoured by those who fall into the 'hierarchist' category. It assumes that the climate system is to a certain degree uncontrollable, but is in fact quite resilient if suitably managed. The risks of climate change are not trivial, but to manage them safely, greater knowledge about the climate system and predictive capability are needed. Such a hierarchist might comment: 'Calculations show that if we can increase the [Earth's] reflectivity by about 3 per cent, the cooling will balance the global warming caused by carbon dioxide in the atmosphere.'

Nature as capricious. This myth is favoured by those who fall into the 'fatalist' category. It sees the climate system as fundamentally unpredictable, influenced by a multiplicity of factors of which humans are but one. Climate has always presented risks to humanity and will continue to do so in the future. Some we will manage, others we won't. Such a fatalist might comment: 'Even if we shut every fossil fuel power station, crushed every car, and grounded every aircraft, the Earth's climate would still continue to change.'

Such categories can, of course, only be approximations for the way in which people think, talk and act in relation to climate change. Cultural Theory has been criticised for being too simplistic to capture complex social and personal realities and for not being rooted sufficiently in empirical evidence drawn across multiple societies. Few individuals will hold to any of these extreme positions, and none will do so consistently. The various categories described here should best be thought of as archetypical 'voices' that are heard in public discourses. At the very least, therefore, the framework points to the multiple influences that people may be exposed to when they evaluate for themselves the significance of the risks of climate change. (We use this framework again later in Chapters 8, 9 and 10 to help us navigate around different 'voices' that speak about climate change.) No matter how strong the consensus may be from scientists about future climate change risks, no matter how convincing may be the 'objective' risk analysis by the experts, the ways in which such risks are perceived, ranked and given priority in people's lives are a function of many things not captured in such narrowly prescribed risk assessments. The knowledge, values and beliefs that we explored in Chapters 3, 4 and 5 are now augmented with combinations of wider cultural and psychological conditioning.

With these insights in place, we can now explore more specifically the ways in which climate change risks are determined, communicated, perceived and believed. We start with the idea of dangerous climate change.

6.3 What Change in Climate is Dangerous: and How Do We Know?

If danger is the state of being exposed or vulnerable to harm or risk, then the idea of 'dangerous climate change' has risk assessment and risk perception at its very heart. The idea of anthropogenic climate change being dangerous was first seriously considered by scientists in the 1950s and 1960s, but it was the 1992 UN Framework Convention on Climate Change that enshrined the idea of dangerous climate change at the heart of climate policy debates. Article 2 states that the ultimate objective of the Convention is 'to avoid dangerous anthropogenic interference with the climate system', an objective that has been widely abbreviated to avoiding dangerous climate change.[13] The idea of danger in the context of global climate change, however, is notoriously difficult to quantify. The problems fall into three groups.

First, framing 'danger' as a result of future (anthropogenic) climate change discounts the dangers that continuously exist for peoples around the world from ongoing and naturally occurring climate risks. It demands that we are able to distinguish between climate risks that are natural and climate risks that are anthropogenic in origin. This distinction is hard to make, since all meteorological risks are now an outcome of a global system which is only semi-natural.

Second, the formulation of Article 2 leaves wide open the question of which categories of danger should count. The metrics for quantifying danger could be dollars (economic damage caused), lives (people killed or disabled), species (measures of biodiversity), or even nations (atoll island states which are inundated and cease to exist). Danger could be determined in the aggregate (for example the total number of lives lost worldwide) or could be constituted as a function of the impact on certain social groups (for example the elderly, the young, the 'innocent'). Very different forms of social justice are implied by these different formulations.

Third, the notion of dangerous climate change in Article 2 was intended as a guide for policy making, but it leaves entirely open the processes to be used to arrive at a suitable and useful definition of dangerous climate change. Is dangerous climate change to be discovered by scientists, analysed by experts, negotiated by governments or voted on by citizens? All four approaches have been used (or at least

---

[13] These concepts are not exactly the same: 'dangerous interference' focuses on the processes of change and 'dangerous climate change' focuses on the outcome of change – but for most purposes of discourse they are interchangeable.

suggested) at different times and in different ways, but each approach raises more questions than it answers.

At the heart of the problem of reaching a position on dangerous climate change is whether one approaches the idea from a top-down or a bottom-up perspective. These two approaches result in what have been called 'external' and 'internal' definitions of risk. 'External definitions [of danger] are usually based on scientific risk analysis performed by experts on the system characteristics of the physical or social world. Internal definitions of danger recognise that to be real, danger has to be either experienced or perceived – it is the individual or collective experience or perception of insecurity or lack of safety that constitutes the danger.'[14]

These two perspectives can lead to very different conclusions about climate risks and what level of climate change might be considered dangerous. From a top-down perspective, climate analysts Brian O'Neill and Michael Oppenheimer proposed an atmospheric carbon dioxide concentration of 450 ppm as a threshold beyond which dangerous climate change would occur.[15] Their specific concerns were to prevent the majority of coral reefs from serious bleaching and to minimise the likelihood of the West Antarctic Ice Sheet disintegrating and the North Atlantic heat conveyor switching off.

In contrast, and from a bottom-up perspective, the residents of some east coast villages in England perceive the rising level of the sea even now to be dangerous. The impacts of the rising ocean are already being experienced for many people living in coastal areas without defence structures. In the village of Happisburgh in south Norfolk, for example, twenty-six houses have recently been abandoned due to eroding soft cliffs and one of the residents expressed the idea of dangerous climate change rather differently from O'Neill and Oppenheimer.

[14] p. 11 in Dessai, S., Adger, N. W., Hulme, M., Köhler, J., Turnpenny, J. and Warren, R. (2004) Defining and experiencing dangerous climate change. *Climatic Change* 64, 11–25.
[15] O'Neill, B. C. and Oppenheimer, M. (2002) Dangerous climate impacts and the Kyoto Protocol. *Science* 296, 1971–2.

'This situation has caused [me] a lot of stress – it is always on my mind. I can't even relax in bed as I'm certain that [the house] will go in the middle of the night.'[16]

Because it is generally easier to conduct formal analyses of dangerous climate change using top-down external definitions of danger, most policy discussions have revolved around the quantification of risks to global-scale climate system and ecosystem functions. However, as the IPCC states in its most recent assessment, 'the definition of dangerous anthropogenic interference cannot be based on scientific arguments alone, but involves other judgements informed by the state of scientific knowledge. No single metric can adequately describe the diversity of key vulnerabilities, nor determine their ranking.'[17]

The idea of dangerous climate change can therefore be viewed from a variety of perspectives. How one arrives at an operational definition of danger involves a mixture of scientific, economic, political, ethical and cultural considerations, among others. As risk analysts Tom Lowe and Irene Lorenzoni demonstrated in an empirical study of the mental maps of twenty-four elite European decision makers on climate change: 'Discourses of danger ... are politically and socially generated, thus entwined strongly with issues of power and morality.'[18] Not only will these considerations differ from society to society, they will also change over time. Risks that were viewed as acceptable in previous generations are no longer tolerated – for example, the health effects of poor air quality in large industrial cities – and novel risks that never

---

[16] Kitty Sinclair, aged 72, quoted 20 October 2004, 'Sea threat to coastal communities', BBC News On-line at http://news.bbc.co.uk/1/hi/uk/3743330.stm [accessed 30 June 2008].

[17] p. 781 in Schneider, S.H., Semenov, S. and Patwardhan, D. (2007) Assessing key vulnerabilities and the risk from climate change, in Parry, M. L., Canziani, O. F., Palutikof, J. P., van der Linden, P. J. and Hanson, C. E. (eds), *Climate Change 2007: impacts, adaptation and vulnerability. Contribution of Working Group II to the Fourth Assessment Report of the IPCC.* Cambridge University Press, pp. 779–810.

[18] p. 143 in Lowe, T.D. and Lorenzoni, I. (2007) Danger is all around: eliciting expert perceptions for managing climate change through a mental models approach. *Global Environmental Change* 17(1), 131–46.

affected previous generations become tolerated – for example the death toll from road transportation.

These contrasting approaches to the definition of danger have also been played out politically. For example, on 11 June 2001, President George W. Bush stated that the emissions targets embodied in the Kyoto Protocol 'were arbitrary and not based upon science' and 'no-one can say with any certainty what constitutes a dangerous level of warming and therefore what level must be avoided'.[19] Yet a few years later UK Prime Minister Tony Blair called upon the world's climate scientists gathering in Exeter to 'identify what level of greenhouse gases in the atmosphere is self-evidently too much'.[20] Different ways of conceiving the idea of dangerous climate change can serve quite different political ends.

The idea of dangerous climate change can only be approached by combining insights from science and insights from social psychology. Danger as envisioned and danger as experienced lead to quite different framings of what it is that needs to be avoided, and both sets of judgements bring values and world-views to the fore. Even in the case of large-scale global threats such as the melting or disintegration of large ice sheets, Oppenheimer recognises that 'natural science alone is not sufficient to demarcate the boundaries of "danger" because evaluation of the uncertainty and timing of disintegration and weighing their relevance [for people] are inescapable aspects of defining danger'.[21]

Interpretations of danger will always be context-specific. Danger only becomes meaningful in relation to particular situations faced by particular societies or individuals at particular times. This tension between the external and internal appreciation of risk is embedded in the notion of dangerous climate change. We need to explore

[19] George W. Bush speech 'President Bush discusses global climate change', 11 June 2001, White House.
[20] Tony Blair speech on climate change, 14 September 2004, London.
[21] p. 1400 in Oppenheimer, M. (2005) Defining dangerous anthropogenic interference: the role of science, the limits of science. *Risk Analysis*, 26(6), 1399–407.

more fully what we understand about perceptions of risk and how these differences may contribute to our disagreements about climate change.

6.4 Perceptions of Climate Risk

External or top-down definitions of climate risk and danger cannot command universal assent for the reasons detailed above. The cultural and psychological contexts of individuals differ. Even if they read or listen to the same scientific description of future climate risks, the way in which those same people receive, process, rank and act on these risks will not be the same. We can further examine the implications of these deep-seated differences in risk perception.

Risk psychologists often distinguish between risks that are 'situated' and risks that are 'un-situated.' This distinction is helpful for us in thinking about the perceptions of climate danger. The risk associated with the introduction of a local waste incinerator – for example, worries about the health of local residents – is a good example of a situated risk. The source of the risk is local and tangible (people can see and smell the incinerator) and this makes it easier to believe – rightly or wrongly – that the local population have a degree of control over the risk (campaigning could get it closed down).

The risk associated with global climate change, on the other hand, is a good example of an un-situated risk. The source of the risk is distant and intangible – no-one can see climate changing or feel it happening – and the causes of the risk are diffuse and hard to situate. Even when an extreme meteorological event occurs and damage is caused, it is not transparent to the victims whether or not it was a risk that could be attributed to anthropogenic climate change. For example, whether the severity of Hurricane Katrina in August 2005 can be attributed to climate change is a matter of dispute among experts, let alone among the public. This characteristic of the risks associated with climate change – the difficulty that individuals have in 'situating' them within their normal daily life experiences – offers plenty of space for disagreement. This may depend, in part, on the different ways of looking at the world captured in the four categories suggested by Cultural Theory (see Figure 6.1). The implications of this ambiguity for the perceptions of climate risks were revealed among a cohort of the American public in a study summarised in Box 6.2.

Box 6.2: How do Americans Perceive Climate Risks?

In a poll of over 22,000 people conducted in twenty-one different countries in 2007 by the BBC World Service,[22] 59 per cent of citizens in the USA believed that it was 'necessary to take major steps very soon' to reduce the magnitude of future climate change. Such a question may be seen as a proxy for expressed 'concern' about future climate risks. Among the surveyed countries, only in South Korea, Egypt, India and Russia was there a lower percentage of citizens calling for such measures. Such generalised polling may be useful for comparing attitudes within and between countries, but by using simple measures of concern (for example, 'how serious a threat is global warming?') such polls offer little insight into the reasons for different public perceptions of climate risks. More interesting are questions concerning the role of scientific and technical descriptions of risk and danger in shaping perceptions of risk, versus cultural or psychological factors. Why do some Americans see climate change as an urgent immediate danger, while others view it as a gradual, incremental problem, or not a problem at all?

---

[22] BBC (2007) Most ready for 'green sacrifices'. BBC World Service survey on climate change attitudes, 9 November 2007, see http://news.bbc.co.uk/1/hi/world/7075759.stm [accessed 30 June 2008].

In a detailed questionnaire-based study of nearly 700 Americans conducted in 2003, psychologist Tony Leiserowitz was able to explore these types of questions about the causes of differential climate risk perceptions.[23] In particular, he examined whether the ideas behind Douglas and Wildavsky's Cultural Theory of risk perception (see Section 6.2) had any empirical backing in the case of climate change. The survey found that the American public possess only moderate risk perceptions of climate change, driven primarily by the perception of danger to geographically distant people, places and non-human nature, rather than by concerns about local climatic risks. Using the categories of Cultural Theory, Leiserowitz found that egalitarians had the highest perception of the risks posed by climate change, while individualists and hierarchists had lower perceptions of climate risks.

Policy preferences were also strongly influenced by these cultural and value categories. Support for national and international climate policies was strongly associated with egalitarian values, while opposition was more strongly associated with individualist and hierarchist positions. These value commitments – these 'ways of life' – were stronger predictors of risk perception than either political party identification (i.e. Democrat or Republican) or political ideology (i.e. liberal or conservative). Risk perceptions were also affected by what Leiserowitz described as 'holistic affect and imagery'. How people reacted to visual images associated with climate change – melting ice, extreme heat, the idea of Nature – was a more powerful predictor of individual risk perception than was their reaction to technical accounts of scientific risk analysis.

These findings match the expectations suggested by Cultural Theory and the associated dominant Nature myths summarised in

---

[23] Leiserowitz, A. P. (2006) Climate change risk perception and policy preferences: the role of affect, imagery and values. *Climatic Change* 77, 45–72.

Box 6.1 – the risks associated with climate change will appear much greater to egalitarians than to either individualists or hierarchists. Risk perception was shown to be greatly influenced by affective and emotional factors, and the survey provided strong evidence that public risk assessments are strongly influenced by experiential processes. As discussed in Section 6.3, definitions of dangerous climate change that rely solely on expert analysis and ignore the emotive and affective reactions of individuals are unlikely to resonate strongly with most people.

Overall, this study revealed how underlying values and world-views strongly condition the way people – in this case members of the American public – think about climate change risks and about the policy options to mitigate global climate change. The perception of climate risks goes well beyond basic issues of scientific literacy, analytical reasoning and technical knowledge. Risk perception and policy preferences are strongly influenced by cultural and personal factors as well. Both individual and social psychology are at work in public risk perceptions of climate change.

As suggested by Douglas and Wildavsky back in 1982, risks are not perceived, assessed and responded to purely on the basis of a universal and 'objective' technical risk analysis. Climate risks are perceived by individuals set within a particular cultural, social and ethical context, in which groups of individuals are predisposed to select, ignore and interpret risk information in different ways. Risk perceptions are thus socially constructed, with different groups prone to take notice of, fear and amplify some risks, while ignoring, discounting or attenuating others. This demonstrates that messages about climate change need to be tailored to the needs and predispositions of particular audiences if they are to resonate with strongly held values. These implications for risk communication are explored further in Chapter 7.

There is another distinction recognised by risk psychologists which also helps us to understand better the diversity of perceptions of the dangers associated with climate change. This is the difference between our *affective* and *analytic* reasoning systems. Our affective evaluation of risks – which in evolutionary terms is an older instinct and one shared with other animals – is intuitive, automatic and fast. It represents risk as a feeling and has a high dread factor. Thus, with an impending tornado bearing down on our house, our instinct is to flee this particular climatic risk.

Our analytic processing system, however, works by the mental processes of assimilation, deliberation and judgement. Requiring reflective thought, it works more slowly than does affective reasoning. The rules of such a process and the appropriateness of use in given situations may have to be taught explicitly, for example to the young, naive or inexperienced. The decision about whether or not to buy a home situated in a known flood plain, for example, would invoke analytic reasoning of the risks involved. Indeed, a decision in favour of doing so might often be referred to in popular discourse as 'taking a calculated risk'.

In many situations, of course, these two reasoning systems are not discrete or compartmentalised and one might override the other. For example, those who engage in the leisure pursuit of bungee-jumping have to overcome their affective reasoning system – 'don't jump off a high platform!' – by deploying the calculus of reflective risk assessment – 'the cord is safe and the procedures well tested'. Conversely, and in the case of climate change, the affective experience of an exceptionally cold summer's day may weaken people's reflective belief in the reality that the world is warming and that, rather than untoward cold, it is in fact the risk of increased heatwaves that needs to be guarded against.

Understanding these simple basics of risk psychology sheds light on the diversity of opinions held by members of the public about the nature and significance of climate change risks. While many people believe that climate change is occurring and that this is a bad thing, the issue isn't as important to them as job security, health care or local environmental issues. The abstract and often statistical nature of the risks associated with climate change lacks the immediacy and situatedness of other risks which *do* evoke a strong visceral reaction. While some people might experience the dangers of weather extremes at first hand – for example, the residents in the Norfolk village of Happisburgh cited above – for many, their primary experience of any dangers that climate change may bring will be a vicarious one. They will either view on their TV screens the (distant) effects of climatic extremes in other parts of the world, or they will be told by expert 'talking heads' on those same TV screens that if we exceed a certain amount of warming the (invisible) North Atlantic heat conveyor will break down. For many in the affluent world, 'climate change affects people only indirectly through rising sea levels, extreme events and the social and economic responses to such effects. While climate change may kill millions, it will be on the death certificate of no-one.' [24]

The above discussion illustrates the many factors influencing the ways in which the risks of climate change are perceived by the public. Some of these factors are linked to the nature of the phenomenon of climate change itself: it is an intangible, un-situated risk. Other factors are the result of the particular contexts that individuals find themselves in: their personal experience of climatic danger; the way their affective and analytical reasonings operate; their placement of trust in experts; their values and world-views.

It is clear that the 'objective' risk assessments performed by expert bodies such as the IPCC do not map unchanged onto the mental models and risk perceptions held by members of the public. There are too many intervening filters, mediators and translators. Furthermore, the time-delayed, ambiguous, remote and often abstract nature of the risks of climate change does not generally evoke strong visceral reactions in the lay public. Their affective reasoning systems are not triggered. As American psychologist Elke Weber describes it: 'The dissociation between the output of the analytic and the affective system may result in less concern than advisable, with analytic considerations suggesting that global warming is a serious concern, but the affective system failing to send an early warning signal.'[25]

This lack of powerful instinctive reactions to the risks associated with climate change has led some scientists, analysts and advocates to find ways of amplifying the risks. We now turn to examine some of the interactions in society which can operate as risk amplifiers and which may get used by different interest groups in their desire to override either the analytic or affective reasoning systems of individual citizens.

6.5 The Social Amplification of Risk

The evidence examined above should be enough to convince us that we cannot simply expect scientific experts to conduct and communicate climate risk assessments and that individuals and social groups will consequently act to reduce those risks. The idea of dangerous climate change requires an appreciation of the experiential (internal) components of risk perception, as well as the more technical (external) approaches to risk assessment. Understanding the role of culture and values in these experiential dimensions of risk perception is also helpful.

But there remains a further complication about how we perceive, evaluate and act on risks such as climate change. We will not understand the full spectrum and complexity of attitudes to climate change risks if we do not appreciate the dynamic processes that are at work in society which can alter both the assessments of risks made by experts

---

[25] p. 105 in Weber, E. U. (2006) Experience-based and description-based perceptions of long-term risk: why global warming does not scare us (yet). *Climatic Change* 77, 103–20.

and the perceptions of risks experienced by individuals. The judgements about which risks become focuses of social concern and political activity, and which risks don't, is not just about the interplay between expert and citizen. It is also a function of the interactions between a much larger number of social actors and institutions, each of whom may (either deliberately or not) amplify or attenuate particular risks. It is this thinking that has led psychologists to develop a framework known as the 'social amplification of risk'; a framework which is helpful for understanding what happens to messages about climate change risks.

The idea of the social amplification of risk, originally developed by American geographer Roger Kasperson and his colleagues, holds that risks become portrayed through various signals – through images, signs and symbols. These signals interact with psychological, institutional or cultural processes in society which can amplify (or attenuate) the perception of risks and their manageability. These processes, or filters, operate through what might be thought of as 'amplification stations' or actors: for example, institutions, social groups or significant individuals. They act and speak so as to transform the signals of risk that are circulating in society.

We can use this framework of thinking to give examples of four different risk amplifiers that are at work in the case of climate change. Of course, none of these amplifiers can easily be isolated from their wider social interactions – which is the whole point of developing a framework within which to think of them – but we can at least demonstrate the variety of risk amplifiers that are in operation.

Our first example is the amplifier of language and metaphor. We will say more about this in Chapter 7: *The Communication of Risk*, but we cite here the coupling of climate change with the language of global terrorism, particularly in the UK print media. In this case the UK Government's chief scientific advisor, Sir David King, acted as the significant individual risk amplifier when, in a well-cited opinion article in the US journal *Science* in January 2004, he stated, 'In my view, climate change is the most severe problem that we are facing

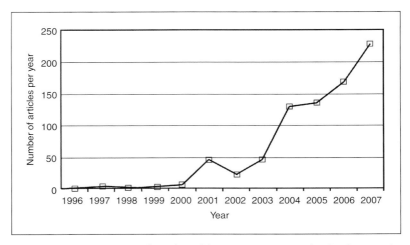

FIGURE 6.3: *Frequency of UK broadsheet newspaper articles for the period 1996–2007 citing 'climate change' and 'terrorism' within the same sentence. Source: Sam Randalls, UCL (personal communication).*

today – more serious even than the threat of terrorism.'[26] In many ways this legitimised the metaphorical comparison of climate change risks with the powerful visceral imagery of terrorism, especially the iconography associated with 9/11, and opened the way for a new linguistic discourse to follow. Evidence for this is seen in Figure 6.3, which shows how the association of terrorism with climate change in UK broadsheet newspapers first peaked in 2001 and then increased almost threefold in 2004 and has continued rising since then.

Climate change is thus benchmarked against a very different type of risk and, through comparisons of other risks with terrorism, becomes triangulated and locked-in with the rise of obesity in Western societies and with the risk of avian influenza (popularly known as 'bird flu'). Thus, 'World governments focus too much on fighting terrorism while obesity and other "lifestyle diseases" are killing millions

[26] King, D. A. (2004) Climate change science: adapt, mitigate or ignore? *Science* 303, 176–7.

more people'[27] and 'Bird flu is now as much of a danger to Britain as terrorism, ministers have been told by the Government's official emergency body'.[28] Rather than evaluate the risks of climate change in their own right, this process of risk amplification articulates climate change risks within the narrative frame of terrorism.

A second example we can give of a risk amplifier within this wider social framework is an institutional one – the international scientific conference on Avoiding Dangerous Climate Change held at Exeter, UK, in February 2005. This conference moved the perception of climate change risks away from prospects of incremental changes in climate, of whatever magnitude, and instead opened up the prospects of large and significant dislocations to the way the climate system worked. Again a powerful metaphor was invoked during the conference to convey this new amplified danger – the idea of 'tipping points or elements'; points of no return. And, as a classic exploitation of the amplification concept, these risks were reported as being 'worse than previously thought', although how much worse was left rather vague.

A third example comes from the institution of the media, which plays a powerful role in the amplification or attenuation of risks. In August 2003 a severe heatwave affected large areas of western Europe and around 15,000 premature additional deaths were reported in France, mostly in the Paris region. The French social scientist Marc Poumadere examined the way in which events unfolded in France during those few weeks in late summer 2003 and how the unfolding heatwave and its impacts were talked about in public, by the Government and by the media. He saw evidence of the amplification of risk at work in the way in which the events were interpreted and narrated. 'A high point of amplification was reached in the quality press when the national daily

[27] Professor Lawrence Gostin speaking at the Oxford Health Alliance Summit, quoted in *The Age*, Melbourne, 25 February 2008.

[28] Bird flu 'as grave a threat a terrorism'. *Independent*, 26 June 2005. See www.independent.co.uk/environment/bird-flu-as-grave-a-threat-as-terrorism-496608.html [accessed 12 November 2008].

Le Monde stated on the front page of its 10 September issue that summer 2003 was the deadliest in France since the end of World War II.'[29]

The study also commented on the contrast with earlier heatwaves that were not subject to the same processes of risk amplification and narration. Thus, in 1976, about 6,000 excess premature deaths were caused in France by a heatwave and yet this event passed completely unnoticed and the death count was only revealed by later retrospective analysis.

A fourth and final example of how the risk amplification process works in society is less specific than the previous three, but it demonstrates the importance of the wider cultural processes at work in the shaping and construction of risk. Sociologist Frank Furedi has written extensively about the emergence of risk cultures in modern societies and links them to wider cultural trends in the West. These cultural moves act as amplifiers of risks by insisting that the world is a far more dangerous place than it was in the past. Furedi has explored this thesis extensively in the context of the fear of terrorism, where he argues that cultural pessimism and a crisis of governance underlie the discourse of 'the war on terror'. But he also draws a wider conclusion, namely that the same underlying cultural trends that enable amplification processes to operate with regard to terrorism, are also at work in society in the case of threats such as climate change.

> The institutionalisation of threat amplification reflects dominant cultural attitudes towards risk and uncertainty. These attitudes have been internalised by officialdom, who are in any case insecure about their own authority. The tendency to expand the territory of the unknown is … an acknowledgement of an absence of cultural and intellectual resources with which to engage with uncertainty.[30]

---

[29] p. 1490 in Poumadere, M., Mays, C., Le Mer, S. and Blong, R. (2005) The 2003 heat wave in France: dangerous climate change here and now. *Risk Analysis* 25(6), 1483–94.

[30] p. 168 in Furedi, F. (2007) *Invitation to terror: the expanding empire of the unknown.* Continuum Books: London.

What we have shown here are examples of some elements of the social amplification of climate change risk; elements which interact across scales, institutions and actors. Far from being able to read off the script of the IPCC assessments what the risks of climate change 'really are', individual citizens find themselves embedded in this tangled web of interactions where risks are symbolised, translated and interpreted in numerous ways and by multiple actors. Some of these amplification (or attenuation) processes are more visible and transparent than others. Thus the risks of climate change communicated by Greenpeace seem quite different from the risks of climate change communicated by, for example, the Ford Motor Company. Other processes, such as the changing cultures of risk that we are caught up in, are more heavily disguised and almost invisible to us. We are left constructing our own mental maps of the risks of climate change, drawing upon our world-views, our personal experiences of climatic dangers, and our implicit hierarchies of trust in the multiple voices we hear and read that are trying to gain our attention.

6.6 Summary

So why do we disagree about climate change? This chapter has explained that one of the reasons we disagree is that we understand and perceive the risks associated with climate change in different ways. Some of these differences arise because of our different outlooks on the natural and social worlds and how we see our relationship with them. Some of these differences arise because we are directly exposed to different levels of climate risk or because we live in different risk cultures. And behind all of this there are the voices in society clamouring for our attention, each seeking, for various reasons, either to amplify or attenuate the risks of climate change which emerge from the processes of negotiated science.

The insights of Cultural Theory suggest that our world-views and our values exert a strong influence on how we perceive climate change and its attendant dangers. Rather than passively being told by experts what the risks of climate change 'really are', and then believing them, many people project their world-views outwards, thereby shaping the sorts of risks associated with climate change in which they are prepared to believe. Someone who views the world's climate as fragile and easily destabilised is more likely to believe intimations that we are approaching a 'tipping point' in relation to ice sheets or ocean currents than is someone who views Nature as benign or tolerant. When scientific assessments clash with deeply held values or outlooks, it may not always be science that triumphs.

Of course, just as scientific assessments of risk may change owing to new knowledge, so too will people's values change over periods of time or even from day to day depending on the particular issue confronting them. People are rarely fully consistent in living out one particular set of values. What our survey has also shown is that apprehending the risks of climate change is a deeply cultural and social activity. If people perceive themselves as vulnerable to climate risks, either by prior experience or through falling into social or geographical categories deemed to be at risk, they are more likely to adopt the high-risk narratives of climate change espoused by some commentators. On the other hand, people living in cultures that are more risk-tolerant, or who fall into social categories that are deemed to have a high coping capacity for climate risks, may well be more sceptical about the quantitative risk assessments that emerge from science and that are subsequently amplified across society.

What does this uneven psychological landscape of climate risk perception imply for questions of urgency, agency and priority about the design and public acceptability of climate change policies? Predicting how different people will perceive the risks of climate change is deeply problematic. There are large differences both within and between different national societies and regional cultures, and these differences may change over time. This offers significant challenges for those seeking to design and implement policy measures that contribute to a goal that is constructed as global in reach (for example, emissions reductions) and that are deemed to be legitimised by the universalising processes of science (for example, limiting global warming to a given magnitude). Equally challenging are the processes we use as a worldwide community to arrive at common and useful definitions of dangerous climate change; definitions which can carry discourses forward towards agreed sets of responses. If perceptions of climate risks are so heterogeneous and cannot be homogenised by universalising IPCC assessments, how is the world to make up its mind about climate change – through an exhibition of power diplomacy between state and non-state actors or through more deliberative and democratic processes? One commentator has suggested that the answer to the question, 'How serious are the risks of climate change?' should be to ask the people – stage a global referendum.[31]

Throughout the above discussion we have seen how central are the media to the ways in which we apprehend climate change risks. Their role as mediators between assessments of climate risks undertaken by scientists, on the one hand, and psychological constructions of climate risks performed by citizens, on the other, performs a powerful function in the understanding of climate change. It is now essential for us to investigate the channels or circuits of climate communication used by the media in greater depth. We need to consider the languages, images and signs that are used in climate change communication, and we need to understand something about the motives and goals of these 'risk amplification stations' existing in our media. We may disagree about climate change because of the multiple and diverging ways in which climate change is talked about and portrayed in our media. This proposition is the subject of Chapter 7: *The Communication of Risk*.

[31] Black, R. (2008) 'Bypassing the blockage of nations', BBC News On-line, 15 January 2008, see http://news.bbc.co.uk/1/hi/sci/tech/7187985.stm [accessed 30 June 2008].

FURTHER READING FOR CHAPTER 6

Douglas, M. and Wildavsky, A. (1982) **Risk and culture; an essay on the selection of technological and environmental dangers.** University of California Press: Berkeley, CA.
Although not specifically about climate change or climate risks, this book remains a classic introduction for understanding the ways in which risks get constructed in society. Douglas and Wildavsky introduce a cultural theory of risk perception and help to explain why some risks become salient and acted on, while other risks are downgraded and ignored.

Leiserowitz, A. A. (2006) Climate change risk perception and policy preferences: the role of affect, imagery and values. **Climatic Change** 77, 45–72.
This empirical study of public attitudes to global warming in the USA tests two ideas: that risk perceptions are influenced by affective imagery and that cultural values influence risk perception. A survey of 670 respondents confirms both hypotheses. Here is empirical evidence for understanding why we disagree about climate change – not to do with the science, and not much to do with cognition, but a lot to do with values, world-views, emotions, trust and experience.

Schellnhuber, H.-J., Cramer, W., Nakicenovic, N., Wigley, T. M. L. and Yohe, G. (eds) (2006) **Avoiding dangerous climate change.** Cambridge University Press.
This is the edited book which resulted from the 2005 Exeter conference on Avoiding Dangerous Climate Change. The book contains forty-one separately authored chapters and, although these are of uneven quality, the book is significant in that it offers, at a particular moment in time, the sum of efforts from natural scientists, and from economists, to address the idea of dangerous climate change.

Thompson, M. and Rayner, S. (1998) Cultural discourses. In: Rayner, S. and Malone, E. L. (eds), **Human choice and climate change. Vol. 1: The societal framework.** Battelle Press: Columbus, OH, pp. 265–344.
This book chapter was prepared as part of the major social science study of climate change completed in the 1990s, and remains one of the best available overviews of how individuals, societies and cultures perceive climate change and the attendant risks. It provides an introduction to Cultural Theory, risk perception, and the role of values and uncertainty in the construction and interpretation of climate risks.

SEVEN

The Communication of Risk

7.1 Introduction

In May 2004, the Hollywood movie *The Day After Tomorrow* went on worldwide release. Tens of millions of cinema goers in over a hundred countries saw the movie which, with gross takings of over $500 million, made a large return from its production costs of $125 million. The film depicts an abrupt and catastrophic transformation of the Earth's climate into a new Ice Age, with North America being in the eye of the cataclysm. It plays upon the scientific uncertainty surrounding a so-called 'tipping point' in the Earth system: the shut-down of the thermohaline circulation (which carries the warm waters of the Gulf Stream into high European latitudes) in the world's oceans. Set against a background of tidal surges, tornadoes, flooding and hurricanes, the human story in the film is about a climatologist who tries to figure out a way to save the world from these cataclysmic changes in climate, while simultaneously trying to rescue his young son stranded in New York, which has been inundated by a giant tsunami and then enveloped in a mega-ice storm.

Climate change is not a usual subject for a Hollywood movie. Although the film makers acknowledged their exaggeration and sensationalisation of the science, they nevertheless claimed that their portrayal of dramatic climate events could have a major influence on the behaviour of society. They suggested that it might motivate people to do something about climate change before it 'became too late'. 'To have a major studio release of a movie tackling a serious issue is a terrific opportunity for Americans to start talking about the reality of the problem [of climate change], what can be done about it and the enormous threat that President Bush is not dealing with.'[1] Producer Mark Gordon hoped his movie would make people think. He stressed it wasn't made to fit any single agenda, but he clearly revelled in the stir that was caused: 'From the box-office point of view, controversy is good. It makes people talk about it ... you couldn't buy this kind of publicity.'[2]

Scientists, politicians, environmental campaigners and their critics also speculated about how the film might impact on public perceptions of climate change and their demand for, or acceptance of, new policy initiatives. Some believed it would increase awareness about climate change and even galvanise the public to effect changes in behaviour and put pressure on governments to act on climate change. Others, however, thought it would reinforce scepticism about climate or else have no impact at all. Advocacy groups around the world prepared for the film's launch by issuing press releases, by distributing leaflets to cinema goers explaining the serious threats posed by climate change, and by organising public meetings and panel discussions about the film's 'serious message'.

But what affect did *The Day After Tomorrow* actually have on cinema goers? Was it an effective way of engaging with a section of the public that might be less easily reached through more conventional forms of science communication? Did it alter the way people viewed

---

[1] Peter Schurman, Moveon.org's executive director, 20 May 2004, Associated Press. www.msnbc.msn.com/id/4900768/ [accessed 3 July 2008].
[2] *Ibid.*, Mark Gordon, producer of *The Day After Tomorrow*.

the science of climate change? Survey work conducted with cinema audiences in a number of countries – the USA, the UK, Germany and Japan – revealed mixed reactions.[3]

Ambiguous and ambivalent indications of attitudinal and behavioural change were revealed among respondents who had viewed the film. Analysis focused on four key social and behavioural issues: people's perception of the *likelihood* of extreme impacts; their *concern* over climate change versus other global problems; their *motivation* to take action; and the locus of *responsibility* for the problem of climate change. Some changes in concern, attitude and motivation were found in a number of viewers. Seeing the film changed some people's attitudes, at least in the short term. These viewers were significantly more concerned not only about climate change, but also about other environmental risks such as biodiversity loss and radioactive waste disposal. While the film increased general anxiety about environmental risks, viewers experienced difficulty in distinguishing science fact from dramatised science fiction. In particular, the dramatic portrayal of climate change in the film reduced viewers' belief in the likelihood of extreme weather events occurring as a result of climate change.

Although the film may have sensitised viewers, and perhaps even motivated some of them to act on climate change, the indications were that the public did not feel they had access to information about what action they could take to mitigate climate change. In addition, the research suggested that any increase in concern about climate change induced by the film appeared short lived, with most viewers treating the film purely as entertainment. Overall, the film sent mixed messages about climate change to viewing audiences and cannot be

[3] See, for example, Leiserowitz, A. A. (2004) Before and after The Day After Tomorrow: a U.S. study of climate change risk perception. *Environment* 46(9), 24–37; Reusswig, F. and Leiserowitz, A. A. (2005) The international impact of The Day After Tomorrow (Commentary). *Environment* 47(3), 41–3; Lowe, T. D., Brown, K., Dessai, S., de Franca Doria, M., Haynes, K. and Vincent, K. (2006) Does tomorrow ever come? Disaster narrative and public perceptions of climate change. *Public Understanding of Science* 15(4), 435–57.

said to have induced the sea-change in public attitudes or behaviour that some advocates had been hoping for.

The Day After Tomorrow is just one of a number of high-profile popular devices for communicating climate change in recent years which have been claimed to have had a powerful effect, either intentionally or unintentionally, on public opinion. In December 2004, just a few months after *The Day After Tomorrow* was released, the best-selling novelist Michael Crichton published his fictional thriller *State of Fear*. This novel portrays the idea of climate change as a conspiracy between scientists and environmental campaigners: a self-important non-governmental organisation deliberately hypes the science of global warming to further the ends of evil eco-terrorists. The conclusion of *State of Fear* is that global warming is a non-problem; it acts as a social attenuator of risk in contrast to the risk amplifiers we saw in the previous chapter. Like *The Day After Tomorrow*, Crichton's novel addresses real scientific issues and controversies, but is similarly selective (and occasionally mistaken) about the basic science.

In May 2006, two years after *The Day After Tomorrow*, another movie was released that has been claimed to have had a more lasting effect on public opinion, especially in America. *An Inconvenient Truth* is a documentary film about global warming, presented by US politician Al Gore, which became the fourth highest-grossing film documentary of all time. The movie had the deliberate goal of awareness-raising and inducing behavioural change[4] and its perceived impact was undoubtedly one of the factors that led to Al Gore being awarded, jointly with the IPCC, the 2007 Nobel Peace Prize. The citation for the Prize stated it was awarded 'for their [Gore's and the IPCC's] efforts to build up and disseminate greater knowledge about

[4] The website for *An Inconvenient Truth* – www.climatecrisis.net – contains specific information and printable tips that are intended to foster individual energy efficiency, including information about how each individual can reduce his or her impact on climate change at home, in commuting, and even nationally and internationally.

man-made climate change, and to lay the foundations for the measures that are needed to counteract such change'.

This chapter examines how the idea of climate change has been represented in a range of media contexts – by scientists, by campaigning organisations, by governments, by advertisers. It discusses the relationship between these representations and the different underlying motivations of the communication agents involved. Public and policy discourses of climate change are certainly influenced by how scientific risks associated with climate change are communicated: the different ways the issue is framed, the different audiences targeted, the different language, stories and visual imagery adopted. We need to understand the ways in which science, policy and the public meet through media-shaped narratives. We need to understand who controls these narratives and the way they influence what people believe about climate change and about its significance. **One of the reasons we disagree about climate change is that we receive multiple and conflicting messages about climate change and we interpret them in different ways.**

We start by considering a number of conceptual models which describe how science might be communicated across and within society (Section 7.2). The traditional 'deficit model' of science communication is no longer tenable; it is not sufficient to argue that more or clearer information about climate change from scientists will lead to greater public engagement with the issue. Neither can it be argued that more scientific *certainty* about future climate change, or better representations of scientific *uncertainty*, will necessarily lead to greater public agreement about what to do in response. There are barriers other than lack of scientific knowledge to changing the status of climate change in the minds of citizens – psychological, emotional and behavioural barriers. We need to understand the complex 'cultural circuits' of science communication in which framing, language, imagery, marketing devices, media norms and agendas all play their part in the construction, mediation and reception of messages.

As we have seen in Chapters 3, 4 and 5, there is no universal way in which science, economics or religion can speak about climate change. When it comes to communicating climate change within the public realm it is therefore no surprise that very different messages about climate change are constructed. The 'framing' of climate change is explored in Section 7.3, showing relationships between how the issue is presented and the intended cognitive or behavioural outcomes. Climate change means different things to different people because we are all exposed to a variety of messages about climate change and we each interpret them according to our own unique background – our knowledge, experiences, values and circumstances. These messages are conveyed through complex combinations of language, images and symbols, some of which are examined in Sections 7.4 (which deals with the use of language) and 7.5 (which examines the use of images). These sections also include case studies about the language of catastrophe and about the challenges of using climate change imagery in environmental campaigning.

Communications experts Craig Trumbo and James Shanahan have researched the ways in which the public understand the ideas surrounding climate change. They observe: 'If public understanding of [climate change] is built on a narrative construct – one subject to a potential fickle storytelling process that can easily be driven in any direction – then politically based policy and regulatory strategies that rely on [such] an authority located in public opinion could be seriously misinformed.'[5] It is the argument of this chapter that the contested nature of the science, economics and beliefs that surround the idea of climate change makes it inevitable that the stories that are told about

[5] p. 203 in Trumbo, C. W. and Shanahan, J. (2000) Social research on climate change: where we have been, where we are, and where we might go. *Public Understanding of Science* 9, 199–204.

climate change in the media *can* 'easily be driven in any direction'. This was one of the ideas we introduced in Chapter 1 – climate as a carrier of ideology – and it is an idea supported by our exploration of the communication of climate change.

7.2 Science Communication Models

A conventional, and still prevalent, view of the relationship between science and the public sees it as consisting of a one-way flow of knowledge and information. Popularised in the 1980s under the slogan 'the public understanding of science', this view requires scientists to become better and more frequent communicators and the public to become more engaged and receptive listeners. The media conduits along which this knowledge passes should be neutral conveyors of scientific ideas and facts. When applied to scientific knowledge that relates to matters of public policy, such as climate change, this model of communication assumes that the provision of scientifically sound information can change public behaviour and increase support for new policy measures. If the public are resistant to these scientific messages, this implies that the public are exhibiting a lack of necessary knowledge – a deficit which needs remedying by science communicators. For obvious reasons, this model of science communication has been termed the 'deficit model'.

Since the late 1990s, a series of seminars organised by the Cambridge Media and Environment Programme has explored the performance of BBC media in reporting a range of environmental issues, most notably climate change and sustainable development. Participants were drawn from the media, government and the world of science. In reflecting on the discussions during these seminars, one of the organisers, British geographer Joe Smith, observed the deficit model in operation through the assumptions brought to the table by the participants. The media had failed in what they viewed as their duty to inform the public. As Smith comments, the scientists 'suggested the media are responsible for public ignorance of both causes and consequences of climate change … in other words, they imagine an uncomplicated flow of [climate change] data from experts, packaged by the media, to an uninformed, receptacle-like society'.[6]

The deficit model of science communication places a very high premium on science for the shaping of individual and collective opinions and adopts a rational-actor model of human behaviour.[7] In relation to climate change it reasons thus: scientific research discovers the problem of climate change; science seeks to identify a range of potential solutions; scientists then inform politicians of these findings and also seek to alter public awareness, attitudes and behaviour by telling them the 'facts' of climate change. We encountered this type of reasoning back in Chapter 3: *The Performance of Science*, when we were examining the role of science in policy. There it was called 'the technocracy model' of decision making.

People and organisations who adopt this mode of reasoning are very likely to end up frustrated. As we have seen in previous chapters, interpretations of climate change by the public are mediated by their values, beliefs and personal experiences, and by cultural norms. This is consistent with the findings of a large survey of American citizens conducted in 2004. Political scientist Paul Kellstedt and colleagues found that 'more scientifically informed' respondents not only felt less personally responsible for global warming, but also showed less concern about it. They concluded that 'the knowledge-deficit

---

[6] p. 1473 in Smith, J. (2005) Dangerous news: media decision making about climate change risk. *Risk Analysis* 25, 1471–82.

[7] The rational-actor model of behaviour assumes that behaviour is motivated by a conscious and rational calculation of advantage, a calculation that in turn is based on an explicit and internally consistent value system. Many social psychologists would disagree that this is an adequate description of how humans appear to behave.

model is inadequate for understanding mass attitudes about scientific controversies'.[8]

The problem is further compounded because the media do not operate as a neutral conveyor of scientific knowledge to a passive audience. They actively and continuously engage in framing, filtering and interpreting messages about climate change using affective and emotive language and imagery. This frustration is well summarised by American geographer Susi Moser, commenting on the deliberations of an expert workshop of climate change scientists, advocates and communicators held in the USA in 2004. 'If only they [the public] understood how severe the problem is ... if only we could explain the science more clearly, train to be better communicators, become more media-savvy, get better press coverage ... The science of global warming is clear – why are we not acting as a society to combat the problem? Why are they not listening? Why is no-one doing anything?'[9]

These frustrations of scientists, campaigners, and perhaps some politicians – their perception of a communication failure – can lead to a variety of reactions. Scientists may be called upon to speak with clearer or more passionate voices; the media may be criticised for balancing scientific views for and against human influence on climate; the public may be blamed for being stubborn and perverse. As risk expert Tom Lowe puts it, 'Risk communicators shout louder to try and shake some sense into people ... The public are on the receiving end of an increasingly distraught alarm call.' [10]

If approaching climate change communication through a deficit model is inadequate for imposing a consensus scientific view of climate

---

[8] p. 122 in Kellstedt, P. M., Zahran, S. and Vedlitz, A. (2008) Personal efficacy, the information environment, and attitudes towards global warming and climate change in the United States. *Risk Analysis* 28(1), 113–26.

[9] p. 3 in Moser, S. and Dilling, L. (eds) (2007) *Creating a climate for change: communicating climate change and facilitating social change.* Cambridge University Press.

[10] Quoted in Revkin, A. (2007) Are words worthless in the climate fight? *New York Times*, 3 December.

change on a passive public, how else might we think of the relationship between science and society? And what do alternative models of science communication imply for the extent to which we can agree or disagree about climate change? One reaction against the deficit model is the idea of dialogue or deliberation between scientists and citizens.

Dialogues imply extended conversations between scientists and the public, between experts and non-experts. These conversations are rooted in the personal experiences and values of the participants as much as they are driven by scientific knowledge. A deliberative approach to climate change communication changes the balance of power between scientist and citizen. Within the context of such extended conversations about climate change, the citizen is able to bring into the open the role of individually held beliefs and values. Citizens are able to see whether the expert knowledge of the scientist is resistant or sympathetic to such personalised viewpoints. Equally, scientists may be able to examine whether the words, numbers or visual devices that they use to convey complex ideas about future climates and their uncertainty find any traction with a listening public. Through dialogues, disagreements about climate change can certainly be aired and different positions understood, even if no resolution to disagreement is forthcoming.

This more engaged form of science communication has been recognised in recent years among many scientific institutions, although it remains far from the norm. For example, a report from the British House of Lords in February 2000 – *Science and Society* – identified the importance of a shift in the culture of science communication from a deficit to a dialogue model. 'Direct dialogue with the public should move from being an optional add-on to science-based policy making and to the activities of research organisations and learned institutions, and should become a normal and integral part of the process.'[11]

---

[11] House of Lords (2000) *Science and society*. Report of the House of Lords Select Committee on Science and Technology, London.

While scientist–citizen dialogues offer a more engaging way to communicate ideas about climate change, by their very design they will always only ever touch a very small proportion of the public. The majority of citizens will not come into direct contact with scientists through extended conversations[12] and will continue to rely upon messages conveyed to them by various forms of media: television, newspapers, websites, magazines, radio. If the media do not, in fact, act as neutral conveyors of information about climate change, we need better ways of conceiving and understanding their role if we are to understand this important communicative dimension of why we disagree about climate change. Instead of the metaphor of linearity in communication implied by the deficit model, a more plausible metaphor that has been proposed is that of cultural circuits and their entanglement.

The idea of 'cultural circuits'[13] offers a very different perspective on science communication from that implied by the deficit model. The latter model sees the senders of climate change knowledge (scientists) only distantly connected to the receivers of such knowledge (the public) through the lines of information flow offered by the media. Scientists, media and public have their own distinct domains and the senders and receivers have little influence on each other. In contrast, a cultural circuits conception of this relationship maintains that both senders and receivers are jointly engaged in shaping and changing the meaning of messages about climate change. The media themselves offer a dynamic arena where these powerful processes are played out. Messages about climate change have no starting point and no ending point; they travel around this circuitry, changing frame, form and meaning as they go.

---

[12] Climate change blogs – such as www.realclimate.org – offer a different means of connecting expert with non-expert, although again this is not mass communication.

[13] The application of the 'circuits of culture' model to climate change studies was made by Carvalho, A. and Burgess, J. (2005) Cultural circuits of climate change in the UK broadsheet newspapers, 1985–2003. *Risk Analysis* 25(6), 1457–70.

This view of the media as a collection of dynamic agents continually active – either implicitly or explicitly – in shaping and reshaping messages about climate change fits much more realistically with what we can observe happening than does the deficit model. It also allows us to understand the role of the media in propagating or even fostering disagreements. We will find examples later in the chapter of how this is achieved through framing the issue of climate change in different ways (Section 7.3), through the use of language (Section 7.4) and through the use of imagery (Section 7.5). For now, let us consider just one example from each of the three domains of this cultural circuitry of climate change communication – the media, the public and science – which illustrates these dynamics at work.

Media. Newspapers continue to operate as a major source of information for the public in many countries, although of slowly diminishing significance in relation to internet-based media and television. A number of studies in recent years have tracked the reporting and representation of climate change in print media in countries such as the USA, the UK, Japan and India. One facet of newspaper reporting that has more rarely been examined is the role of ideology[14] in print media representations of climate change. Newspapers, either through their proprietors or through tradition, frequently espouse specific ideological positions.

Communication scientist Anna Carvalho explored the relationships between the ideologies of three national UK newspapers – *The Times*, the *Guardian* and the *Independent* – and their coverage and representation of climate change during the period 1985–2001.[15] She

---

[14] As we defined it in Chapter 1, an ideology may be thought of as a 'body of doctrine, myth or belief that guides an individual, a social movement, institution, class, or large group'. In this case, we are thinking of ideology as guiding the way in which the world is idealised and reported by a newspaper.

[15] Carvalho, A. (2007) Ideological cultures and media discourses on scientific knowledge: re-reading news on climate change. *Public Understanding of Science* 16, 223–43.

found significant differences between coverage in *The Times* and
the other two newspapers which she attributed in part to the more
conservative ideological slant of that newspaper. The *Guardian* and
the *Independent* gave greater weight to scientific assessments of climate change risks and demanded stronger political intervention. *The
Times*, however, favoured a greater emphasis on scientific uncertainties about future risks and adopted a more liberal, market-oriented
view of potential policy options. Similar differences in the reporting
of climate change can be linked to the different ideological slants of
national American newspapers; for example, the *Wall Street Journal*
(more sceptical of climate change) compared with the *New York
Times* (more accepting of the potential risks and the need for government intervention). Newspaper media therefore actively shape stories
about climate change, whether stories of the scientific risks involved
or stories about possible policy solutions. Science, ideology and public
audiences become entangled in the web of these cultural circuits.

Public. New media forms and technologies are altering the ways in
which citizens interact with 'news' and science. They are offering
platforms for more diverse public voices to be heard on issues such as
climate change. New televisual and web-based diary styles, for example, allow a new depth to public communication about everyday lives;
a phenomenon that Richard Sambrook – director of BBC Global
News – has termed '360 degree storytelling'. Far from being passive
receivers of expert science, media communication now allows citizens
to actively challenge and reshape science, or even to constitute the
very process of scientific investigation through mass participation in
simulation experiments such as 'climateprediction.net'.[16] New media
developments are fragmenting audiences and diluting the authority
of the traditional institutions of science and politics, creating many

---

[16] This project harnesses the power of tens of thousands of home computers
around the world to generate computer predictions of future climate. See www.
climateprediction.net/ [accessed 14 November 2008].

new spaces in the twenty-first century 'agora' (places of public assembly) where disputation and disagreement are aired. As geographer and media analyst Joe Smith has remarked, television 'programming that blurs the boundaries between news/current affairs and other broadcast categories … offers further opportunities for engaging publics in understanding and debating climate change risks'.[17]

Science. The internet is still a relatively new medium and its roles in the circuits of communication are still being formed and described. The internet has revolutionised the ways in which communication of scientific ideas occur, but has also at times blurred and confused the role that science plays in public debates. A good example of this occurred in November 2007, when the results from 'an important new study' were posted on the internet. Referring to a new scientific paper in the *Journal of Geoclimatic Studies*, the post suggested that undersea bacteria were mostly responsible for the build-up of carbon dioxide in the atmosphere and *not* fossil fuel emissions from human energy generation. Within a few hours, maybe even minutes, this new scientific finding was appearing on blogs and circulating around the world on email lists, finding a particularly receptive audience among groups and commentators who were sceptical of the human influence on climate.

The story was of course a hoax, with the perpetrator being British novelist and journalist David Thorpe. He later explained his reasons for creating such a spectacularly successful hoax:

> Sometimes fiction and satire can reach places facts alone can't – in the right context … What the hoax showed is that there are many people willing to jump on anything that supports their argument, whether it's true or not. What we wanted to emphasise is that it's necessary to achieve scientific validity using the peer-review model. Proper climate science makes every attempt to do this, and is a constantly evolving and self-refining process, as all science is.[18]

---

[17] Smith, Dangerous news, p. 1480.

[18] *New York Times* science blog: http://dotearth.blogs.nytimes.com/2007/11/11/the-life-and-death-of-a-climate-hoax/ [posted by Andy Revkin, 11 November 2007].

What this example shows is the dynamic interplay between science, media and public at work through the medium of the internet. As *New York Times* science correspondent Andy Revkin commented, it showed 'the amazing power of the Web to amplify and then dismantle fictions at light speed'. The power of the media to shape, convey and dismantle 'truth' is ever-changing.

We saw in Chapter 3: *The Performance of Science* the difficulty of retaining a view of science in which an unambiguous truth is spoken to power, especially science that addresses public policy issues such as climate change. We now see similar difficulties in a view of science communication in which knowledge about climate change is transmitted unambiguously to the public. The aphorism 'all publicity is good publicity' may hold true to some extent for politicians or celebrities seeking to retain a public profile. It is less clear whether it applies to complex issues such as climate change. 'Raising awareness' of climate change can never be neutral or an end in itself. Each story we hear about climate change is framed in a certain way to emphasise a particular facet of the phenomenon, whether this be concerning the severity or the responsibility for the problem or the options and responsibility for implementing solutions.

7.3 No Message is Neutral

If we understand the relationships between science, media and public in the ways implied by the metaphor of tangled circuitry, we can begin to understand the multiple ways in which an issue such as climate change can be framed. Rather than there being only 'facts' about climate change proclaimed by institutions such as the IPCC – 'facts' received intact by the masses – the circuitry of the media offers spaces and creative potential for social actors to filter, amplify and rhetoricise these 'facts' in multiple ways. As Carvalho and Burgess explain in the context of newspapers: 'Different social actors are locked in competition around how climate change risk is to be framed in the media … their framings are always mediated through each newspaper's preferred ideological world-view.'[19] No message about climate change is neutral; certain aspects of the story are emphasised and other aspects are downplayed or ignored.

The idea of 'frames' and 'framing' has emerged in the social psychology community over recent decades and has been particularly applied to how news, ideas and issues are reported in the media. Put simply, 'Frames organise central ideas, defining a controversy to resonate with core values and assumptions … They allow citizens to rapidly identify why an issue matters, who might be responsible and what should be done.'[20] One can use this idea of framing to examine, from a number of different standpoints, how ideas about climate change are communicated: the originator of a news story (e.g. the scientist or campaigner), the transmitter (e.g. the journalist or media institution), and the audience (e.g. the public or policy makers). Originators of stories may frame them according to their specific world-view or according to their understanding of the world-views of the audience(s) they are trying to reach. The public on the other hand, faced with a daily torrent of competing or conflicting news stories, will often use their own framing of issues as a way of filtering or selecting stories that accord with these frames. And the media will also be active through their role in re-framing issues according to their own ideologies, norms or audience preferences. This is known in psychology as the 'confirmation bias'.

When looked at in this way, climate change offers opportunities for governments, organisations and individuals to adopt an almost unlimited variety of framing devices. One can trace the changing dominant frames that have been used to present climate change

[19] Carvalho and Burgess, Cultural circuits of climate change, p. 1458.
[20] p. 56 in Nisbet, M.C. and Mooney, C. (2007) Framing science. *Science* 316, 56.

at the largest scale, that of international institutions. In the 1980s, climate change first came to global prominence as an environmental issue and hence attracted the attention of environmental campaigning organisations such as Greenpeace and Friends of the Earth. The IPCC was a creation of environmental bureaucrats (UN Environment Programme) and meteorologists (World Meteorological Organization) rather than one of, for example, development economists (World Bank). Over the subsequent twenty years, however, one can see how different meta-frames for climate change have been adopted: as a development issue (the prominence of climate change at the 1992 Earth Summit and the 2002 Johannesburg follow-on), as an economic issue (the 2006 Stern Review), as an issue of national and global security (the UN Security Council debate of April 2007; see Chapter 9: *The Way We Govern*), or as an issue of morality and social justice (the World Council of Churches; see Chapter 5: *The Things We Believe*).

Each of these framings of climate change emphasises certain aspects of the issue, while de-emphasising others. They also carry implications for how the causes of climate change are portrayed – over-consumption by the North, a failure of markets, structural injustices in the world economy – how urgent responses to climate change should be, and who should be the main agents responsible for implementing solutions. Framing climate change as a failure of markets, for example, implies that it is market entrepreneurs, economists and businesses that need to take the lead in 'correcting' this failure. Framing climate change as a challenge to individual and corporate morality, on the other hand, suggests that very different cohorts of actors should be mobilised.

None of these ways of framing climate change can be claimed to be wrong in any absolute sense. Equally, none of them offer a 360-degree view of climate change. By definition, frames select and emphasise certain facets of an issue and must therefore de-select and de-emphasise others. Communicating climate change can therefore never be merely 'raising awareness' or simply presenting 'the facts'. Raising awareness of what aspect of climate change? Raising awareness with what audience and end in sight? In the very act of constructing and communicating a story about climate change, certain causes, actors or responses are favoured or demonised.

A good example of contrasting media framing of climate change comes from the newspaper reporting of the IPCC's Fourth Assessment Report and the publication of the Working Group II Report on Impacts, Adaptation and Vulnerability. On Saturday 7 April 2007, the day after the report was finalised and approved by governments in Brussels, nearly all of the national British daily newspapers carried one or more stories on the findings. Drawing upon exactly the same IPCC press release and the same IPCC Summary for Policymakers, the reporting in two of the more popular newspapers – the *Daily Mirror* and the *Daily Express* – was especially noteworthy. The *Daily Express* ran a headline 'Melting Ice Doomsday' and introduced their item with 'Britain faces catastrophe within 50 years unless governments act now to prevent climate change'.[21] In contrast, the *Daily Mirror*'s headline was 'Boom & Dust', with the following introduction: 'Britain would become warm and prosperous through global warming but it will bring disaster to poor countries, the world's top scientists warned yesterday.' [22] In one frame, climate change brings catastrophe to Britain, a prospect which the newspaper uses to challenge national governments to act. In the other frame, climate change offers economic benefits to Britain, while countries in the developing world reap disaster.

Different framings resonate more powerfully with some audiences than they do with others. Mike Shanahan, press officer at the International Institute for Environment and Development in London,

---

[21] *Daily Express*, 7 April 2007, see www.express.co.uk/posts/view/3796 [accessed 14 November 2008].

[22] *Daily Mirror*, 7 April 2007, see www.highbeam.com/doc/1G1-161683981.html [accessed 14 November 2008].

TABLE 7.1: *Examples of ways of framing climate change, and the audiences most engaged.*

Climate change frame	Audience engaged
Scientific uncertainty frame	Those who don't want to change
National security frame	As above, but now inspired to act
Polar bear frame	Wildlife lovers
Money frame	Politicians and the private sector
Catastrophe frame	Those who are worried about the future
Justice and equity frame	Those with strong ethical leanings

Source: Shanahan (2007).

illustrates six different framings of climate change,[23] each constructed to appeal to particular audiences (the implication being that they will have less – or no – appeal to other audiences). These frames, together with their prospective audiences, are shown in Table 7.1. We can see how framings may relate, albeit unevenly, to the different 'ways of life' presented in Chapter 6: *The Things We Fear* through the lens of Cultural Theory (Box 6.1). A catastrophe frame, for example, may engage those who see Nature as ephemeral (egalitarians), whereas the scientific uncertainty frame may engage those who see Nature as capricious (fatalists).

Communicating climate change always carries a message or, to put it more formally, any communication about climate change will always frame the issue in a particular way; for example as scientists seeking to reduce uncertainty, as social justice movements seeking to protect the vulnerable, or as carbon entrepreneurs seeking to make money through new commodity markets. Using the idea of frames to help us understand why we disagree about climate change requires

[23] Shanahan, M. (2007) Talking about a revolution: climate change and the media. *COP13 Briefing and Opinion Papers*, IIED: London.

us to take one further step: to understand the roles of language and imagery in the construction of these different framings. This is the step that the next two sections of this chapter help us to make.

7.4 Linguistic Repertoires

Closely allied to the idea of frames are what have been called linguistic repertoires, 'routinely used systems of language for describing and evaluating actions, events and people'.[24] A repertoire might include a distinctive vocabulary, a set of stylistic features, certain metaphors, idioms or images. We might well expect different ways of framing climate change to be related in some way to the adoption or preference for certain linguistic repertoires. Framing climate change as a security issue, for example, might favour the adoption of phrases such as 'coastal defence' and 'climate protection' or metaphors such as climate change as 'a weapon of mass destruction'[25] and waging a 'war on climate change'. Using language is never neutral, but always active. Tracing and following linguistic repertoires associated with climate change is one more way in which we may reveal how and why we disagree about climate change.

This approach has been pursued most systematically in two studies (*Warm Words* and *Warm Words II*)[26] conducted for the London-based Institute for Public Policy Research (IPPR). These studies examined articles from UK newspapers and magazines, as well as TV and radio news clips and adverts, press ads and websites. Over 600 such items were studied in each of two periods – from November

---

[24] Segnit, N. and Ereaut, G. (2007) *Warm words II: how the climate story is evolving* IPPR/Energy Savings Trust: London.

[25] Houghton, J. T. (2003) Global warming is now a weapon of mass destruction. *Guardian*, 28 July 2003.

[26] Ereaut, G. and Segnit, N. (2006) *Warm words: how are we telling the climate story and can we tell it better?* IPPR: London; and Segnit, N. and Ereaut, G. (2007) *Warm words II: how the climate story is evolving.* IPPR/Energy Savings Trust: London.

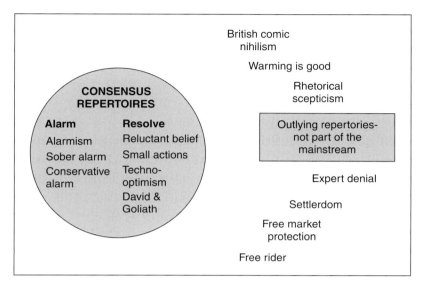

FIGURE 7.1: *Consensus and outlying 'linguistic repertoires' used in public discourses around climate change in UK society between March and July 2007. Source: Segnit and Ereaut (2007).*

2005 to February 2006 and again between March and July 2007. The research concluded in July 2006 that, 'the climate change discourse in the UK today looks confusing, contradictory and chaotic'. The authors interpreted twelve linguistic repertoires that captured this diversity, offering labels ranging from 'alarmism' and 'sober alarm' to 'techo-optimism' and 'rhetorical scepticism'. Eighteen months later, the repeat analysis suggested that some convergence was occurring around a smaller cluster of repertoires that embraced both 'alarm' and 'resolve' about climate change (Figure 7.1). These repertoires might collectively be considered as constituting the mainstream consensus, while several outlying repertoires representing a variety of minority and more sceptical narratives continued to circulate.

These conditions of linguistic diversity relate only to one particular society and present snapshots taken at two moments in time; many more comparative cross-cultural and longitudinal studies would need to be conducted to build up a more comprehensive insight into how the language of climate change evolves across societies and through time. The real value of these two IPPR studies for our current purpose, however, is their revelation of the diversity of linguistic repertoires of climate change that can co-exist in a society at the same time. Each repertoire reveals something different about the multiple and perhaps overlapping constituencies that tend to use them. The 'reluctant belief' repertoire – a pragmatic, if weary, acceptance that climate change is real and that something needs to be done – suggests a very different outlook regarding the urgency and motivation to respond to climate change than that implied by the 'David and Goliath' repertoire – aggressive, oppositional campaigning environmentalism and radical heroism. An example of contention about the validity of the language adopted by the 'alarmist' repertoire is summarised in Box 7.1.

Box 7.1: The Language of Catastrophe

One of the linguistic repertoires suggested by the *Warm Words* studies of the IPPR (see main text) was labelled 'alarmism', caricatured by the aphorism 'we're all going to die'. Central to this repertoire is the use of words such as 'catastrophe', as in the example, 'scientists are increasingly concerned at the possibility of abrupt catastrophic climate change'.[27] A number of scientists have challenged the aptness of this language for describing what science has discovered about climate change and how science projects future climate risks. In an article for BBC News On-line in November 2006, Mike Hulme asked the question, 'Why is it not just campaigners, but politicians and scientists too, who are openly confusing the language of fear,

---

[27] p. 13 in McDermott, T. (2007) No time to lose. *New Statesman*, London, 29 January.

terror and disaster with the observable physical reality of climate change, actively ignoring the careful hedging which surrounds science's predictions?'[28] In challenging the widespread adoption of this linguistic rhetoric, Hulme concluded, 'The IPCC scenarios of future climate change … are significant enough without invoking catastrophe and chaos as unguided weapons with which forlornly to threaten society into behavioural change.'

The use by scientists of the language of catastrophe was subsequently defended by another climate scientist, Australian modeller James Risbey. In his essay, Risbey claims that the language of catastrophe is a suitable, indeed a necessary, means of communicating the scientific nature of the problem. He distinguishes between discourses which are 'alarmist' (rhetorical, inconsistent with the science and implying fatalism) and those which are 'alarming' (emerge from current scientific understanding and intended to alert the public of the need to change course).[29] Risbey goes on to accuse those who do not adopt such urgent language in their descriptions of the science as failing in their civic duty to inform the public, a 'scientific reticence' which falls short of the standards of impartial communication.

Risbey's response to Hulme's argument encapsulates several of the dilemmas, several of the reasons we disagree about climate change, already encountered in this book. The different positions adopted by these two experts on the most suitable language used to communicate climate change can be traced back to several facets we have already examined: the relationship between consensus and uncertainty in science; the role of science in public policy; the extent to which beliefs and values insinuate themselves in scientific judgements; the different views of Nature as suggested by Cultural Theory.

[28] Hulme, M. (2006) 'Chaotic world of climate truth', BBC News On-line, 4 November 2006, http://news.bbc.co.uk/1/hi/sci/tech/6115644.stm [accessed 3 July 2008].

[29] Risbey, J. S. (2008) The new climate discourse: alarmist or alarming? *Global Environmental Change* 18(1), 26–37.

There are many ways in which language plays a central role in the communication of climate change and, through the different choice of words, opens up spaces for misunderstanding, if not disagreement. We briefly examine here just two more of these: the adoption of the descriptors 'climate change' or 'global warming' to describe the phenomenon of concern; and the ways in which scientific uncertainty about future climate is captured through words.

In the English language, the terms most often used to describe the physical transformation of global climate through human modification of the atmosphere have varied over time. The 'greenhouse effect' or 'enhanced greenhouse effect' were terms widely used in the early scientific framing of the issue in the 1980s and early 1990s (see Chapter 2: *The Discovery of Climate Change*). These have subsequently been largely replaced either by the more generic term 'climate change' or the more evocative expression 'global warming'. More recently, the adjective 'catastrophic' has increasingly been used in descriptions of the phenomenon, thus 'catastrophic climate change' now circulates as a separate descriptor, carrying a greater sense of urgency than the more neutral 'climate change'.

Each of these terms not only conveys differences in technical understanding of what is being described, but also has different impacts on lay audiences. In a detailed examination of public understanding of the terms 'climate change' and 'global warming', social psychologist Lorraine Whitmarsh found different associations were triggered by these two terms among a cohort of the British public. 'Climate change' evoked a lower level of concern than did 'global warming', the latter being more commonly associated with heat-related impacts and, erroneously, stratospheric ozone depletion. Conversely, while it has been claimed that the term 'catastrophic climate change' (see Box 7.1) should be adopted in order to alarm the public,[30] positive

---

[30] For example, see Read, R. (2007) Emergency talk. *Guardian*, 13 November: '"Climate change" is a criminally vague and anodyne term that is dangerous for us to use. Let's not fool ourselves by using warm words such as "climate change" (or indeed "global warming", which still to my ears sounds pretty misleadingly

messages tend to be more attractive and effective in motivating behaviour change than negative ones. As Whitmarsh concludes, in this context, even if nothing else is signified by this study, 'there needs to be an explicit recognition that terminology is not neutral and should not be used indiscriminately'.[31]

The second example of how our use of language can exacerbate disagreement comes from a more technical aspect of science communication – how scientific uncertainties are represented. We saw in Chapter 3: *The Performance of Science* that there are a number of ways in which scientific assessments, such as the IPCC, handle uncertainty about future climate projections. In particular we drew a distinction between model-based and expert-based disagreement (see Box 3.1). These different approaches to capturing scientific uncertainty may have much wider ramifications for how the public or policy makers interpret scientific information about possible future outcomes. Risk analyst Tony Patt explored whether the portrayal and communication of scientific uncertainty made any difference to people's belief in scientific projections or their willingness to favour specific policy actions on climate change. He concluded that whether uncertainty was presented as arising from disagreements 'between models' or disagreements 'between experts' – in other words, the social history of that uncertainty – did indeed matter. 'The fact that conflict has arisen about particular estimates of the future may signal features not only of the science, but also of the politics of that science, that are relevant for policy makers to learn [from].'[32]

pleasant. I meet lots of people … who say things like, "Yeah, we could use a little global warming around here!"). Talking instead about averting "climate catastrophe" is not alarmism. It is simply calling things by their true names.'

[31] p. 18 in Whitmarsh, L. (2008) What's in a name? Commonalities and differences in public understanding of 'climate change' and 'global warming'. *Public Understanding of Science* doi:10.1177/0963662506073088.

[32] p. 45 in Patt, A. (2007) Assessing model-based and conflict-based uncertainty. *Global Environmental Change* 17(1), 37–46.

The language we use in telling stories about climate change may reflect our disagreements about the saliency and urgency of the issue while, at the same time, careless and unthinking use of our language in this respect may perpetuate or exacerbate these disagreements. These dual characteristics of language can also be found in the iconography of climate change, to which we now turn.

7.5 The Iconography of Climate Change

Climate change has long presented challenges for the media, campaigners and scientists wishing to use visual devices in their communication to lay audiences. The standard scientific index for monitoring and predicting global climate change is the globally averaged near-surface air temperature. Yet this quantity – a disembodied global temperature – can be neither directly measured, seen nor photographed. As media expert Stuart Allan and colleagues have remarked about a whole genre of technologically induced risks, including climate change: 'They operate outside the capacity of (unaided) human perception. This im/materiality gives [such] risks an air of unreality until the moment they materialise as symptoms. In other words, without visual presences, the hazards associated with these technologies are difficult to represent as risks.'[33]

Making climate change visual – through drawings and paintings, photographs and icons – has a long history, whether constructed for artistic, historical or normative purposes. The late-medieval winter landscapes by Dutch Renaissance painter Pieter Bruegel – for example his 1565 'Hunters in the Snow' – remain evocative depictions of an era when the European climate was colder than today, as do depictions by wood-engravers and artists of London frost fairs on the River Thames from the seventeenth to nineteenth centuries (Figure 7.2).

[33] p. 3 in Allan, S., Adam, B. and Carter, C. (eds) (2000) *Environmental risks and the media*. Routledge: London.

FIGURE 7.2: An 1844 engraving from the Illustrated London News of the
last Thames frost fair of February 1814.
Source: National Maritime Museum, London.

Glaciers and palm trees have been enduring motifs of climate change,
as demonstrated by the Swiss geographer Stefan Brönnimann in his
survey of over a hundred years of print media.[34] And, more recently,
the Cape Farewell project has sought to motivate international artists
and writers to find innovative ways of representing the implications
of climate change through taking them on dedicated voyages to the
High Arctic. The aim is to make the invisible global climate, visible
and local in new ways. 'We intend to communicate through artworks
our understanding of the changing climate on a human scale, so our
individual lives can have meaning in what is a global problem.'[35]

[34] Brönnimann, S. (2002) Picturing climate change. *Climate Research* 22, 87–95.
[35] David Buckland, artist and project director of Cape Farewell, quoted in Martin,
    C. (2006) Artists on a mission. *Nature* 441, 578.

We may find creative ways beyond the computer graphics and maps of science for visualising climate change and what it means to different people. But are we increasing shared awareness and inspiring common advocacy among the public, or are we simply reflecting the growing range of meanings that climate change has for society? In other words, does the iconography of climate change help us to agree about what is signified by the phenomenon or does it entrench, in visual form, the reasons for our disagreement? There is certainly a wide spectrum of images that are used by communicators as icons or signifiers of climate change. The case study of climate change campaigns operated by Greenpeace over recent years (Box 7.2) is one good example of this. Other examples include the adoption of the polar bear as a 'poster-child' of climate change and the ways in which climate change has been signified by commercial advertisers in recent years. We will briefly examine both of these examples.

Box 7.2: Greenpeace and the Picturing of Climate Change

The environmental campaigning organisation Greenpeace first attended to the idea of global climate change as a priority issue in the late 1980s. Photography is central to the organisation's campaigning philosophy – 'bearing witness to environmental damage' – and it has successfully used visual imagery in campaigns against whaling, marine pollution and nuclear power. Yet climate change has presented Greenpeace with a major dilemma – how does one photograph a phenomenon that is largely invisible and one whose most dramatic effects are scheduled for the as yet unrealised future?

Media expert Julie Doyle has traced the ways in which Greenpeace has pictured climate change in its campaigning literature over the period 1994–2005.[36] This short history of climate iconography from within one organisation powerfully reveals the different ways in which the visual has been enlisted in support of different framings of climate change. Greenpeace has been consistent in communicating climate change to worldwide audiences and campaigning against it, and yet very different messages have been conveyed through their adoption of different visual icons.

Doyle identifies five phases in the visual iconography of climate change used by Greenpeace: climate change as time bomb, as fossil fuel addiction, as catalyst for renewable energy, as dirty politics and, finally, as 'here and now' reality. For each of these five framings of climate change, different visual icons have been emphasised in Greenpeace's marketing campaigns. This is not so much revealing disagreements about climate change, but revealing the multiple ways in which climate change can be framed and communicated to different effect at different times.

With their 1994 report *Climate Time Bomb*, Greenpeace sought to drive home the message that the future impacts of climate change would be on a par with a nuclear holocaust. The cover image was of a sunset (or sunrise), edited to resemble the mushroom cloud of a nuclear warhead explosion (see Figure 7.3). In 1997, their campaign moved on to draw attention to the underlying cause of anthropogenic climate change – the dependency of the world economy on fossil fuels. The report *Putting the Lid on Fossil Fuels*, published in 1997, carried a stark image of a red, overheating Earth, to drive home the message of the global consequences of continued fossil fuel consumption. This contrasted with the more upbeat framing of climate change from 1998 onwards, in which the opportunities for a major switch away from fossil fuels to renewables was emphasised. Thus the report *New Power for Britain* in 1998 displayed

---

[36] Doyle, J. (2007) Picturing the clima(c)tic: Greenpeace and the representational politics of climate change communication. *Science as Culture* 16(2), 129–50.

FIGURE 7.3: *Front cover image of the Greenpeace report* Climate Time Bomb, *published June 1994.*
Source: Greenpeace (2004).

wind, sun and waves as the energy saviours for avoiding the worst that climate change might throw at us.

By 2002, Greenpeace's climate change frame had changed again, this time to attack the Bush Administration and ExxonMobil as twin targets that were obstructing progress on tackling climate change. Their 'Stop Esso' campaign from 2001/2 featured the face of George W. Bush adjacent to an adulterated Esso logo. As Doyle commented, climate change was by now perceived by Greenpeace 'to have developed sufficient cultural resonance for the public to understand [the image] without including evidence of climate [change] impacts'.[37]

The final phase of visual framing explored by Doyle is the more recent one from the 2000s in which visual imagery is used by Greenpeace to document the past and present effects of global warming on the physical environment, notably on ice sheets and glaciers. This visual device is consistent with Greenpeace's corporate philosophy of 'bearing witness to environmental damage', but the juxtaposed imagery of glaciers from a hundred years ago and from today also evokes nostalgia for an irretrievable past. This is quite a different emotive pull from their original 1994 image of a 'nuclear' time-bomb, with the emotional connotations of danger, fear and death.

In each of these framings, climate change is represented through iconography which is aimed at inducing very different reactions in audiences. We might all recognise that climate change is the common thread through this story, but what climate change signifies in each case is quite different, as are the responses thereby provoked in Greenpeace's target audiences.

Polar bears. Polar bears are frequently used as an iconic species in the communication of climate change by the media; photographs of polar bears commonly adorn newspaper articles and campaigning tracts in the Western press. During protests in July 2001 at the Sixth Conference of the Parties to the Framework Convention in The Hague, environmental campaigners dressed up as polar bears and dramatised 'die-ins' to drive home their point about climate change threatening this iconic species. Popular articles frequently suggest a rapid and alarming decline in polar bear populations under climate change. Underlying such practices lies the view that, 'the polar bear is coming to symbolise the disappearing north, the end of the kind of climate we all grew up with'.[38]

[38] p. 426 in Slocum, R. (2004) Polar bears and energy-efficient light-bulbs: strategies to bring climate change home. *Environment & Planning D: Society and Space* 22, 413–38.

But the iconography of the polar bear as signifier of climate change may as easily be seen as a site of heightened controversy and disagreement than as one that effectively stabilises debate and concern around a shared natural 'treasure'. Survey work shows that the power of the polar bear icon to represent climate change in the minds of the public rests on its emotional appeal. While powerful among those to whom polar bears matter, this has little or no traction among those who are not interested in polar bears. This visual representation of climate change is also fragile because the dynamics of polar bear populations are contested among ecologists. To what extent are polar bears seriously threatened by climate change over and above other human pressures on their habitats? Danish statistician Bjørn Lomborg opened his recent book *Cool It* by questioning whether polar bears were indeed the 'canaries in the coalmine' they are frequently portrayed to be. 'Once you look at the supporting data the narrative falls apart.'[39]

Advertising climate change. One of the most powerful cultural uses of visual imagery is in commercial advertising. Value is added to products by using evocative images and texts, often appropriating meanings from one cultural context and attaching them to products in unrelated settings to advance sales (commercial advertising) or behavioural change (social marketing). Here we find more examples of where the iconography of climate change has solidified differences between climate change discourses rather than acted as a common visual denominator to induce agreement.

American public health expert Stephen Linder has examined how commercial advertisers have taken the idea of climate change, and the

[39] p. 2 in Lomborg, B. (2007) *Cool it: the skeptical environmentalist's guide to global warming.* Cyan-Marshall Cavendish: London. For a considered investigation into the future fate of polar bears see: O'Neill, S., Osborn, T. J., Hulme, M., Lorenzoni, I. and Watkinson, A. R. (2008) Using expert knowledge to assess uncertainties in future polar bear populations under climate change. *The Journal of Applied Ecology* 45(6), 1649–59.

presumed widespread public concern about it and, through visual and textual cues, used climate change to position their product or company in the market-place. On the one hand, companies such as the Ford Motor Company and Ben & Jerry's ice-cream draw upon themes of corporate social responsibility to make their products appear responsible. A Ford advert from May 2001 therefore shows the iconic image of the sun rising over the orb of the Earth with the slogan, 'Global warming. There, we've said it. Some find our stand on global warming unique – mostly due to the fact that we actually have one.'[40] Ben & Jerry's assume that their audience know about climate change and want to do something to stop it, and thus associate their product indirectly with this assumed desire. Over their advertising image of a tub of ice-cream is the slogan: 'Ben & Jerry's ice cream: help put the freeze on global warming.'[41]

On the other hand, advertisers may also invert the presumed social conscience about climate change. Using text and imagery in an ironic way, they promote *greater* consumption, the opposite outcome from those seeking behavioural change through social marketing. Climate change is hence parodied in an attempt to lure world-weary and cynical customers. Linder gives the example of Ben Sherman Menswear who overlay an image of a rugged and resistant individual with the slogan, 'If the effects of global warming are everywhere, why's my flat so cold?'[42] And another instance of a television advert for Australian Foster's Ice beer takes the parody of climate change one step further. The basic message of the advert is to combat global warming by consuming Foster's Ice beer to stay cool.

> The advert begins in mock serious tones, with some of the most cli-
> chéed images of global warming reviewed in quick succession. The light-
> ing is very dim, but the images feature fire and natural disaster. The

[40] Cited on p. 120, Linder, S. H. (2006) Cashing-in on risk claims: on the for-profit inversion of signifiers for 'global warming'. *Social Semiotics* 16(1), 103–32.

[41] *Ibid.*, p. 122.

[42] *Ibid.*, p. 127.

voiceover projects an ominous sounding authority; the effect is disorienting. Finally, there is an empty set with a globe. The voiceover reviews the conventional set of personal mitigation options, familiar tips for reducing greenhouse gases by changing one's consumption habits. Then comes the punch line: a final option added is '… to say bollocks to it and enjoy an ice cold Foster's …'. The globe collapses into ashes, and the scene cuts away to a kangaroo tapping its foot to 1920s jazz by an inviting pool in an oasis-like setting, drinking a Foster's Ice. The kangaroo offers a knowing wink to complete the parody. The final scene overlays the concluding message in bold print: 'Global Cooling'.[43]

In these examples, climate change – and the signs, symbols and images adopted by commercial advertisers – becomes just one more cultural idiom for promoting products and consumption. Adverts which parody the idea of climate change, such as Foster's Ice, reinforce contrary messages about climate change by referring to advertising efforts built around global warming and then mocking or subverting them.

7.6 Summary

So why do we disagree about climate change? This chapter has suggested that one of the reasons is that we frame, narrate, picture and interpret climate change in quite different ways. Messages about climate change emerge from all parts of the complex cultural circuitry which shapes news, prioritises stories and conveys ideas around our societies. Consensus scientific announcements about the prospects for future climate change or crafted statements about desirable policy frameworks issuing from meetings of the Parties to the Framework Convention rapidly change shape and gain new ideological baggage as they circulate through these digital networks. Messages about climate change become increasingly divergent, often confusing and

---

[43] *Ibid.*, pp. 125–6.

sometimes conflicting, as they are heard, translated and re-assimilated by the world's 6.8 billion inhabitants. It is not long before novelist Ian McEwan's apt metaphor becomes reality, 'Can we agree among ourselves about climate change? We are a clever but quarrelsome species – in our public discourses we can sound like a rookery in full throat.' [44]

Climate change is framed in a multitude of different ways, either informed by the world-views of those communicating or filtered by the intuitive world-views of those listening. Different climate change discourses use different linguistic repertoires, often related to the specific goals of the discourse coalition involved. The existence of different visual representations of climate change – including the different wider iconographies of climate change – loads attitudinal or behavioural outcomes in favour of preselected conditions. At the same time, trends in new media practices, most notably through digital and internet technologies, are opening the way for even greater fragmentation, liberalisation and democratisation of social discourses about a whole range of public policy issues. They may empower more participatory approaches to debating scientific knowledge and designing social responses to climate change, but it seems unlikely that, by themselves, these new media environments will either promote or enable greater agreement about climate change. In the communication of climate change, it seems that the centripetal forces at work are stronger than the centrifugal ones.

We have explored in this chapter the many ways in which climate change can be portrayed in our media, through different combinations of frames, language and imagery, and how these portrayals change meaning as they circulate. One framing of climate change that has saliency because of its appeal to our ethical instincts for justice and equality is one that claims that 'climate change will hit hardest on the poor'. Developing countries are frequently portrayed in this

[44] p. 3 in British Council (2005) *Talking about climate change*. British Council: Manchester, UK.

frame as needing the help of the developed world to cope with the consequences of climate change. Visual images are frequently used by development organisations such as Christian Aid, which depict flood victims in the global South with textual overlays calling out to Northern readers; for example, 'Do us a favour will you? Write to your MP about that Climate Change Bill.'[45] Are these expressions of a campaigning moral outrage or are they manifestations of a smug paternalism or an imperialising form of neo-colonialism? They certainly engage a different social class from that targeted by Foster's Ice beer.

We may disagree about climate change because of the contrasting ways in which we approach the questions of global inequality and sustainable development. It is to these dimensions of climate change that we now turn.

FURTHER READING FOR CHAPTER 7

Ereaut, G. and Segnit, N. (2006) **Warm words: how are we telling the climate story and can we tell it better?** Institute for Public Policy Research (IPPR): London.

Segnit, N. and Ereaut, G. (2007) **Warm words II: how the climate story is evolving.** Institute for Public Policy Research (IPPR)/Energy Savings Trust: London.
These two reports from an independent think-tank examine the ways in which climate change is being discussed in the public domain in the UK. It illustrates a range of different linguistic repertoires that are commonly found, each representing a different conception of climate change as a 'problem' and the options available for 'solutions'. The 2006 survey was followed by an updated account the following year.

Killingsworth, M. J. and Palmer, J. S. (1996) Millennial ecology: the apocalyptic narrative from *Silent Spring* to global warming. In: Herndl, C. G. and Brown, S. C. (eds), **Green culture: environmental rhetoric in contemporary America.** University of Wisconsin Press: Madison, WI, pp. 21–45.
This is an excellent essay which explores thirty-five years of environmental rhetoric from the 1960s to the 1990s starting with Rachel Carson's Silent Spring and ending with global warming. Although it obviously misses the story of the last ten years, it places the language and discourse of environmental (climatic) apocalypse into a longer historical context. The

---

[45] For example, advertisement in the *Guardian*, 29 September 2007.

enduring point is made that the aim of rhetorical apocalyptic warnings is not to predict the future, but to change it. Such discourses are always oppositional, iconoclastic and faith-based, rather than calculative and prediction-based.

Moser, S. and Dilling, L. (eds) (2007) **Creating a climate for change: communicating climate change and facilitating social change.** Cambridge University Press.

This edited volume is the most comprehensive book yet published about the communication of climate change. It collects together over thirty different contributions from (mostly American) academic, civic and governmental perspectives which examine the challenges of communicating climate change to facilitate societal response. The short chapters include a mix of descriptive, critical and practical perspectives on these challenges.

Smith, J. (ed.) (2000) **The daily globe: environmental change, the public and the media.** Earthscan: London.

This book offers an accessible account of the state of knowledge about media treatment and public understanding around the world of key environmental issues such as climate change. It incorporates a wealth of expertise and insight from distinguished journalists, politicians, researchers and environmentalists. The book offers practical examples of successful new forms of communication and lays the foundation for effective strategies aimed at informing public debate on the choices and challenges presented by global climate and environmental change.

EIGHT

The Challenges of Development

8.1 Introduction

In March 1987, the Brundtland Commission submitted its report on strategies for securing sustainable world development to the United Nations General Assembly. Four years in the making, *Our Common Future* by Gro Harlem Brundtland and her commissioners offered the world a diagnosis of the ills facing humanity and laid out a vision for building a future which was 'more prosperous, just and secure'. The central idea of *Our Common Future* was that of sustainable development, a framework for integrating environmental policies and development strategies. Thus:

> sustainable development seeks to meet the needs and aspirations of the present, without compromising the ability to meet those of the future. Far from requiring the cessation of economic growth, it recognises that the problems of poverty and underdevelopment cannot be solved unless we have a new era of growth in which countries of the global South play a large role and reap large benefits.[1]

---

[1] p. 43 in World Commission on Environment and Development (1987) *Our common future*. Oxford University Press.

Anthropogenic climate change was a relatively low-key issue in 1987 (see Chapter 2: *The Discovery of Climate Change*), emerging only slowly from its scientific heartland and still the subject of quite specialist and elite discourses. Yet *Our Common Future* addressed nearly all of the dilemmas that climate change has brought into sharper focus over the intervening twenty years. The language of sustainable development has been increasingly deployed in debates about climate change; for example the interdependence between the environmental, economic and social dimensions of global change; bringing the well-being of future generations forward into debates about the responsibilities of the present generation; and paying attention to the institutional (in-)adequacies of a globalising world. These are all now common strands of climate change discourse, and all three were presaged by the Brundtland Commission.

Despite this rhetorical power and conceptual elegance, putting the idea of sustainable development to work in operational policy settings has proved much harder. We might all endorse the principles outlined in *Our Common Future*, but we do not have a common understanding of what it means in practice. We use the idea of sustainability and make it work for us in a multitude of different and often competing ways: we now have categories of actors such as sustainable consumers, sustainable tourists, sustainable farmers, sustainable house-builders, sustainable foresters, sustainable entertainers. But sustainable development has been critiqued for its lack of attention to power structures (for example what about the role of the military or of elite financiers?), for its anthropomorphism (what about the needs of non-humans?), and for its marginalisation of the poorest (what about the voice of the property-less?). And we have ended up with almost tribally polarised positions. As one commentator puts it: 'Can ecological economists wailing about the folly of unrestrained resource depletion and capitalist cheerleaders ballyhooing the booming economy, possibly have the same thing in mind when they talk of sustainable development?'[2]

At the heart of this confusion are the different ways in which we think about progress. One version of this story emphasises the following. Life expectancy for most people has increased substantially over the past 200 years. World economic activity has doubled every twenty to thirty years, raising per capita incomes for the large majority. The ingenuity and creativity of a growing global population has offered many of us unrecognisably more diverse and stimulating cultural experiences than enjoyed by previous generations. And we are exposed to much greater ethnic and cultural diversity in our societies than has ever been the case before. Isn't it better to be alive in the twenty-first century than in any previous one?

Yet there is another version of this story that can be told. Premature infant deaths in their tens of millions per year are still occurring; deaths due to malaria, malnutrition and poverty, which are entirely preventable. There remain over a billion people in the world whose spending power is less than the purchasing equivalent of a dollar per day. The global level of consumption exceeds the renewable resources of the planet by a factor of more than two. Incidents of genocide, terrorism and structural racism cast shadows over our hopes for the future. And even in the most advanced economies, psychological illnesses related to poor mental and dietary health have been increasing for several decades.

What then is progress – and are we progressing? Does the idea of climate change help or hinder our visions of the future? For some campaigning agencies, the answer is clear: 'Climate change is already a deadly reality that poses more of a threat to development than any other single issue. Floods, famine, drought, disease and conflict are wreaking havoc in poor countries and are set to do untold damage

---

[2] Meyercord, K. (2006) *Sustainable development: an oxymoron?* ZeroGrowth: a new ethos for the millennium. See www.zerogrowth.org/sustaindev.html [accessed 7 July 2008].

as climate change accelerates.'[3] And, for others, climate change appears as a matter of fundamental justice, where the world is polarised between the perpetrators and the victims: 'One group of people (namely rich people everywhere, but mostly in rich countries) have caused the problem [of climate change], and another group of people (namely poor people especially in poor countries) will suffer most of the adverse consequences in the near term.'[4]

This is a common framing for climate change, but is it helpful? Is the most relevant way of redressing past injustices or the best way of tackling poverty to start with climate change? Does the idea of sustainable development have anything to contribute to our search for agreed and universal responses to climate change? Or does the idea of climate change have anything to contribute to our ideas of human development that is sustainable?

Our belief in the idea of 'progress' and our views of development are powerful shapers of our attitudes to climate change. Our definition of poverty and how we encounter and interpret the inequalities in the world are also important. This chapter outlines some of these different beliefs, views and definitions, and explains why an understanding of climate change cannot be separated from an understanding of development. **One of the reasons we disagree about climate change is because we understand development differently.**

We start, in Section 8.2, by answering the question: What do we mean by development? Several different development paradigms have emerged in the second half of the twentieth century: post-War colonial welfare and development; neo-liberalism and free trade; sustainable development (both reformist and radical variants); and, more recently,

[3] Publicity statement from Christian Aid at the launch of their report *Climate of Poverty*, 15 May 2006, London.
[4] Huq, S. and Toulmin, C. (2006) *Three eras of climate change*. Sustainable Development Opinion Papers, IIED: London.

what has been called 'post-development'. Alternative visions of the future – expressed, for example, through the outworking of economic principles or the securing of social justice – reveal differences between these development paradigms. These visions of development have different implications for how we approach climate change.

However we conceive development, the world remains deeply unequal. Increasingly divergent life chances exist between the 'bottom billion' and the rest, while multiple barriers remain for many of those who aspire to basic standards of well-being and personal security. The 2002 Johannesburg Summit articulated many of these contours of inequality and also agreed a set of development goals and institutional innovations to secure them. We are more than half-way to the target date of 2015 and yet the Millennium Development Goals remain an unforgiving pull on the conscience of the world. In Section 8.3 we therefore ask the questions: Within this context of enduring poverty, what significance does climate change have? Is climate change the primary threat to securing sustainable development for the 3 billion people denied it today and for the additional 2–3 billion due to arrive on the planet in the next fifty years? Or are other threats to development – debt burdens, poor governance, trade tariffs, existing climate risks – more significant than climate change? We find very different answers, depending on where we look.

Unavoidable in these considerations is the question of population and demography. Section 8.4 examines the relationship between development, population and climate change and asks why we find it so difficult to bring ourselves to debate the role of population in climate change. If there is a 'safe' level of climate change, is there not also a desirable world population? And, if so, how would we know?

We also reveal a number of other contentious issues which highlight some of the underlying tensions between securing development and tackling climate change (Section 8.5). Proposed 'solutions' to climate change emanating from developed nations look very different to a citizen or policy maker in Kampala or Dacca than they do to an observer in Washington or Sydney. For many, securing reliable and safe energy is more important than reducing its carbon content; access to global markets through the removal of trade barriers and enhanced material exchange is more important than counting up the food miles; and growing the international tourism economy matters more than the carbon dioxide emitted by the air carriers. And the 'new' development agenda – adaptation to climate change, now beginning to be resourced from embryonic climate change funds – might be a distraction from the more fundamental development rights of poverty alleviation, basic health care and education.

8.2 What Do We Mean By Development?

The arrival in the 1980s of the idea of sustainable development was a reaction to a growing recognition that the previous three decades of international development had placed inordinate stresses on the world's ecosystems. As first examined by the World3 modelling team in the 1972 *Limits to Growth* report to the Club of Rome (see Box 8.1), five major trends were of global concern: accelerating industrialisation, rapid population growth, widespread malnutrition, depletion of non-renewable resources, and a deteriorating environment. Yet the model of development pursued in the years after the Second World War had focused primarily on the goal of economic growth. It was pursued initially by the fading colonial powers of Europe and then through the 1960s and beyond by the newly empowered national and international development agencies operating under Cold War rivalry. The social goals of development – alleviating poverty and reducing inequality – were to be achieved primarily through improving the global standard of living. Growth in national gross domestic product (GDP) and in personal income became the criterion of success.

Measured against this criterion, this model of development had considerable success through the 1950s, 1960s and 1970s. The old liberal narrative of human progress still had some traction, empowered by the 'white heat of technology', by the 'Green Revolution' and by generally increasing life expectancies. Yet, by the 1980s, the optimism driving forward this narrative was being challenged. There was a growing reaction against this model of development because it was perceived as being a surrogate for neo-colonial Westernisation and because the inadequacy of using GDP as the measure of progress was being exposed. We have already commented (in Chapter 4: *The Endowment of Value*) on the limitations of GDP as an index of economic performance in the context of climate change policy assessments. The wider charges brought against it in the 1980s were that GDP tells us nothing at all about ecological well-being and, as we have seen in Chapter 5: *The Things We Believe*, it captures only a limited set of the dimensions of social and political life which are needed to describe human well-being.

It was against this background of rising dissatisfaction that the Brundtland Commission reported in 1987 and offered a new set of conceptual tools with which to reinvent the ideas of development and progress. Sustainability was the goal, and sustainable development was the process of moving towards that goal. Rather than using GDP as the indicator of progress, a new set of measuring concepts were needed. Two of these are shown in Figure 8.1, which plots the development of the world and of the large continental regions between 1975 and 2003 according to two measures of sustainability: the human development index (HDI) and the ratio of a region's ecological footprint to the global biocapacity. A high value of the former (measuring human well-being beyond narrow economic criteria) and a low value of the latter (measuring, on an inverse scale, ecological well-being) are desirable. No world region has managed to reach and remain within the sustainable quadrant of the diagram and, for the more advanced regions of the world, each increment in human well-being has only been achieved at a proportionately increasing loss in ecological well-being.

The analysis shown in Figure 8.1 – and others like it – takes seriously the conceptual lessons of the Brundtland Commission. Development

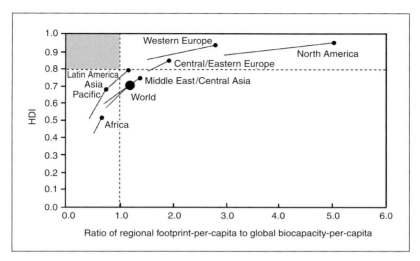

FIGURE 8.1: *Global and regional trends in sustainable development. Regional and world development (HDI: human development index) versus resource demand (footprint-to-global biocapacity ratio). Points indicate values for 2003, and grey trailing lines show trends from 1975 to 2003. The shaded box represents a domain where both points meet one suggested minimum criterion for sustainable development, i.e. HDI ≥ 0.8, and Footprint-to-Global biocapacity ratio ≤ 1.0.*
Source: Moran et al. (2008).

involves more than simply maximising economic growth as measured by GDP. But, as the goal of sustainability seems as far away as ever, some have argued whether the ideas and concepts behind sustainable development are still useful. Commenting on the last twenty years, economist Chris Sneddon and colleagues observe, 'The primary drivers of environmental degradation – energy and material use – have burgeoned and inequalities in access to economic opportunities have dramatically increased within and between most societies. Why then revisit an effort that ... has been so overwhelmed by history?'[5]

Such disillusionment may not be universal, or even typical, but if sustainability is the desirable destination for humanity and if

[5] p. 254 in Sneddon, C., Howarth, R.B. and Norgaard, R.B. (2006) Sustainable development in a post-Brundtland world. *Ecological Economics* 57(2), 253–68.

sustainable development is the journey to reach it, there seem to be many different vehicles that have been constructed for making the journey. And with climate change being presented as one of the gravest threats to the achievement of sustainability, it is important that we understand what some of these constructions look like. The family of sustainable development discourses can be grouped into those that look more 'reformist' and those that look more 'radical'.[6] We look at these two groupings below and offer an example of an approach to climate change policy making that would be consistent with each position.

Reformist movements

Sustainable development demands that ecological sustainability is reconciled with economic growth and the procurement of social goals. One obvious way of achieving this in a market-dominated economy is to adequately price the goods and services offered by ecosystems, including the climate. This is the idea of 'market environmentalism', a means of achieving sustainability which has been powerfully used in debates about climate change policy. If climate change is the result of the 'single biggest market failure' (cf. the Stern Review; see Chapter 4), then the solution is to correct the market by pricing carbon (and other greenhouse gases) explicitly, allowing it to trade carbon out of the energy mix. Market environmentalism sees 'green growth' as the route to poverty alleviation and to restoring the health of ecosystems and the stability of climate.

This thinking lies behind moves to price the 'climate services' offered by tropical rainforests, thus providing a market-based incentive for tropical countries to enhance forest conservation. New 'green' investment companies, such as Canopy Capital, see sustainable development very much in these terms. On the expectation that the global

---

[6] Grist, N. (2008) Positioning climate change in sustainable development discourse. *Journal of International Development* 20(6), 783–803.

market will in due course recognise a monetary value to the services offered by the rainforests, in March 2008 the company agreed to make payments to the government of Guyana for the protection of 370,000 hectares of Iwokrama rainforest. The return on this investment is anticipated to be a share of the future rights to the natural functions of the forest – storing carbon, generating rain and offering a habitat for wildlife.[7] This is a classic example of speculative global capital seeking out high-risk, but potentially lucrative, business ventures.

A second reformist interpretation of the pathway towards sustainability would take a more sceptical view of the efficacy of the market. The idea of 'ecological modernisation', originally espoused by Dutch political scientist Martin Hajer in the 1990s, describes sustainable development as being secured through more direct regulation of various industrial and consumption practices. Rather than focusing on greening or correcting the market alone, ecological modernisation seeks to 'green' technology and to reduce wasteful consumption. It has a much stronger managerial focus than does market environmentalism and seeks strategic alliances between state, business and civil society to provide a political environment conducive to the flourishing of technological modernisation.

Sustainable development as ecological modernisation offers a different prescription for addressing climate change. Climate stability will be secured through policy measures that promote energy efficiency, that increase resource recycling, and which stimulate technological innovation, especially with regard to energy generation. This approach to sustainability would endorse the Clean Development Mechanism (see Section 8.5) as a means of ensuring that environmental goals are secured in developing countries using Northern capital. It would also support the development of technology for carbon capture and storage – for instance, extracting carbon dioxide from the flue gases of coal-fired power stations and burying it underground.

---

[7] Anon. (2008) Racing to hug those trees. *Economist*, London, 27 March.

A third reformist approach to thinking about sustainable development focuses attention on individual consumers and their role in social networks and movements. This pathway might be called 'environmental populism' and gains some of its legitimacy from movements such as Local Agenda 21 (brought into existence during the 1992 Rio Earth Summit) and various rights-based and participatory approaches to development. The argument here is that sustainable development cannot be secured either by the market or by state-sponsored interventions to regulate polluting and wasteful production and consumption practices. Without a powerful and effective groundswell of popular opinion among citizens – whether those of the North, those of the South or, preferably, the citizens of both – all efforts at securing sustainable development will be stillborn.

In the context of climate change, environmental populism implies new social forms of co-operation and participation for securing climate goals. Thus the non-governmental organisation Practical Action works with local people to develop more flood-resistant homes in vulnerable delta communities in Bangladesh, mobilising local environmental knowledge and local materials to ensure sustainability. And, in a different context and culture, Carbon Rationing Action Groups (CRAGs) – local community groups to support and encourage citizens to downsize their personal carbon footprints towards a globally equitable and sustainable level – have emerged in recent years in the UK and some other OECD countries.[8]

Radical movements

Market environmentalism, ecological modernisation and environmental populism may be offered as three pathways to sustainability and securing climate stability which demand reform, but not overthrow, of the prevailing neo-liberal capitalist political economy.

---

[8] See, for example, Carbon Rationing Action Groups, www.carbonrationing.org.uk [accessed 7 July 2008].

These reformist approaches are well captured by the title of Jonathan Porritt's 2005 book, *Capitalism: As if the world matters*, and also by the rhetoric of new business movements such as 'base' (Business and a Sustainable Environment). 'Contributing to the global shift to a low carbon economy is a business imperative and not an option. It opens a new landscape of opportunity and will generate major new alignments of business relationships. 'base' will help companies of all sizes make those new alignments ... sustainability and profit can co-exist.'[9]

But there are also radical critiques of these reformist pathways to sustainability, critiques which challenge the ability of reformist approaches to be effective in the light of the urgency, complexity and scale of the task of reducing global carbon emissions. Radical approaches to sustainable development imply very different approaches to analysing the root problems that drive climate change and hence very different approaches to designing climate policy. We briefly mention three of these.

Ever since Thomas Malthus' 1798 *Essay on the Principle of Population*, there have been resurgent movements which have emphasised limits to the Earth's resources and how many people the planet can support. Some more recent variants of neo-Malthusianism are explored in Box 8.1, as well as some of the political implications of adopting a strong limits-based approach to the question of climate change. This discourse of non-negotiable limits, applied to climate change, lies behind many of the rallying calls which have been heard in recent years to 'act now before it's too late' and 'ten years to save the world'. While this diagnosis of approaching climate breakdown does not itself offer unique prescriptions for securing sustainable development, it unites disparate movements drawn from science, environmental campaigning and certain political elites in their desire and demand

---

[9] Tom Burke, founder and Board chairman of Business and a Sustainable Environment (base), [see www.businessandasustainableenvironment.com/], *base newsletter*, London, April 2008.

for action – urgent action – to be taken to avert the impending catastrophe. It is the urgency of such action, whatever form it takes, that warrants the 'radical' label.

Box 8.1: Climate Change and the Discourse of Limits

The discourse of environmental limits was made most salient through the publication in 1972 of *The Limits to Growth* by the Club of Rome, an elite think-tank founded in 1968 by Aurelio Peccei and Alexander King. The key message of the report was that global economic growth could not continue indefinitely because of the limited availability of natural resources. 'If the present growth trends … continue unchanged, the limits to growth on this planet will be reached sometime within the next hundred years. The most probable result will be a rather sudden and uncontrollable decline in both population and industrial capacity.'[10] Subsequent revisions of this report have been issued, most recently in 2004, attending to earlier criticisms about underestimating the roles of technology and efficiency savings to expand these limits. The core message remained only slightly modified: 'We worry that current policies will produce global overshoot and collapse.'[11]

Many commentators have remarked on the similarity in underlying philosophy between *The Limits to Growth* and Malthus's 1798 essay on limits to perpetual population expansion: 'Population must always be kept down to the level of subsistence.' But the idea of environmental limits to human activity emerges in many other places in

---

[10] p. 1 in Meadows, D.H. (1972) *The limits to growth: a report to the Club of Rome.* Universe Books: New York.
[11] Meadows, D.H., Randers, J. and Meadows, D.L. (2004) *Limits to growth: the thirty-year update.* Earthscan: London.

ecological thinking. It is closely linked to the idea of carrying capacity, which many population ecologists have used and demonstrated in different ecosystem contexts. Carrying capacity is also related to the concept of the ecological footprint – the aggregate area of productive land functionally required to support any given human population. This lies behind the recent claims that, while a current American lifestyle applied to the present world population would require five planet Earths to sustain it, a Sierra Leonean lifestyle applied to all would only require half a planet (Figure 8.2).

Related to the idea of environmental limits is the idea of thresholds. Beyond some threshold of extraction, consumption or pollution, the functioning of an ecosystem or an element of an

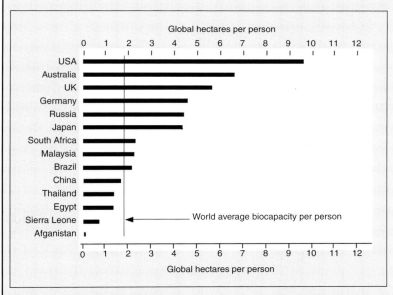

FIGURE 8.2: *Ecological footprint for selected countries expressed as global hectares per person. The world's biocapacity is estimated as just under 2 hectares per person. Data relate to 2003.*
Source: Redrawn from World Wide Fund for Nature (WWF, 2008).

ecosystem will be critically impaired, perhaps irreversibly. Thus, in discussions about 'acid rain' in the 1980s, the notion of 'critical loads' was used to set policy targets for allowable levels of sulphur emissions in Europe. Once the critical load of sulphur dioxide in the atmosphere was exceeded, unacceptable damage was deemed to occur. This notion lies behind most of the air quality standards now in place for urban environments.

These various neo-Malthusian ideas and concepts have now been applied to climate change, lending legitimacy to discourses about sustainability which emphasise the immediacy and urgency of the problem. As we examined in Chapter 3: *The Performance of Science*, the adoption by the UN Framework Convention on Climate Change of the concept of 'dangerous anthropogenic interference with the climate system' has attracted policy analysis to deliberate on what should be a desirable limit of climate perturbation. The EU states the limit to be 2°C above pre-industrial global temperature, although this temperature change translates only vaguely to a greenhouse gas concentration target. A global temperature as a limit not to be exceeded also allows the mobilisation of the concept of thresholds; levels of change beyond which climate system functions may break down and/or become irreversible. This concept, in recent years, has attracted the metaphor of 'tipping points', or 'tipping elements', in the Earth system.[12]

The discourse of limits – whether applied to resource use, to ecological footprints or to dangerous functional thresholds – offers a particular way of approaching the idea of sustainability. Yet the essence of this debate should not be about forensically and objectively trying to establish what such limits are, but rather about understanding to what extent limits are substitutable or adaptable.

[12] See Lenton, T. M., Held, H., Krieglar, E., Hall, J. W., Lucht, W., Rahmstorf, S. and Schellnhuber, H. -J. (2008) Tipping elements in the Earth's climate system. *Proceedings of the National Academy of Sciences* 105(6), 1786–93.

Neoclassical economics, for example, is based on a belief that the factors of production are, to a great extent, substitutable for one another – if one resource is depleted or if one ecosystem is degraded, then alternatives will either be found elsewhere or invented. Neo-Malthusianists, however, would contend that substitutability through technological innovation itself has limits – at some point humanity's capacity for adaptation will lead to an economic, ecological or cultural crash. Pathways to sustainability therefore seek to avoid this point.

Closely allied to the framing of climate change in neo-Malthusian terms of irreversible limits is what has been called 'the eco-anarchist movement'. Anarchistic ideas have influenced direct action groups around the world, including Greenpeace, Earth First! and the Chipko movement. Eco-anarchists demand radical solutions to what they see as the ultimate problem of climate change; namely the current world capitalist order, unbridled consumption and excessive materialism. The radical UK climate campaign, Rising Tide, explains climate change thus: '[It] is a direct result of the economic domination of Northern interests and transnational corporations. We call for climate justice through solutions that address structural inequalities and recognise the historical responsibility of the rich nations for the problem … we call for an immediate end to oil exploration and a dismantling of the fossil fuel economy.'[13]

Sustainable development in this eco-anarchist view is not about trading carbon, nor about pouring tens of billions of dollars into clean technology, nor even about mobilising the might of the green consumer. 'No matter how many low-energy light bulbs you install, or

[13] Rising Tide (2007) *Rising Tide UK: taking action on the root causes of climate change*, Principles 1 and 3. See http://risingtide.org.uk/about/political [accessed 8 July 2008].

how much recycling you do, there is still the need for more systemic changes to take place in society.'[14] In this radical view of sustainable development, the entire growth paradigm is challenged – whether conventional economic growth or sustainable green growth. The World Bank, the Stern Review and the Brundtland Commission are all advocating unsustainable economies and hence perpetuating climate change.

Understanding these different conceptions of the pathways to sustainability is crucial for our understanding of why we disagree about what we should do about climate change. We see here further evidence of the different 'ways of life' introduced in Chapter 6: *The Things We Fear*, using the idea of Cultural Theory, and further examined in Chapter 7: *The Communication of Risk*, through the idea of preferred linguistic repertoires. Without precisely mapping the sustainable development discourses above onto the four cultural categories we introduced in Box 6.1, we can nevertheless see at least some degree of correspondence. Individualists and egalitarians might well sit comfortably with eco-anarchists and the ephemeral view of Nature implied by neo-Malthusianists; whereas hierarchists are more likely to adopt market environmentalism or ecological modernisation as their creed for sustainability.

With this foundation of understanding, we now need to examine more closely the ways in which these different attitudes to development affect the prognosis of what should be done about climate change. We start by looking at the relationship between climate change and poverty.

---

[14] Smith, D. M. (2006) *Just one planet: taking action on poverty and climate change.* Practical Action: Rugby, UK.

8.3 Climate Change and Poverty

When discussing either climate change or poverty, the numbers of people claimed to be most disadvantaged are frequently measured in billions. Paul Collier's recent book, *The Bottom Billion: Why the poorest countries are failing and what can be done about it*,[15] focuses on the poorest billion people on the planet, many of them in Africa; while the first of the UN's 2015 Millennium Development Goals is to halve the near billion people in the world 'who go to bed hungry every night'. With climate change we hear the statistics presented in similar terms. 'Billions of people face shortages of food and water and increased risk of flooding, experts at a major climate change conference have warned.'[16]

The numbers may be similar in magnitude, but the people cited are not the same people. Most of the climate change billions are claimed to be people of the future – either today's children turned adults by 2050, or else the children of the future, as yet unborn. The billions cited by Paul Collier and the Millennium Development Goals, however, are people alive in the world today. And this brings us to one of the hard-edged questions that circulates through the arguments about climate change and sustainable development: are we doing as much for today's poor billions as we are seeking to do for tomorrow's vulnerable billions? Or inverting the question and using the slogans of two recent mass movements: are we doing as much to make poverty history as we are doing to stop climate chaos?[17]

[15] Collier, P. (2007) *The bottom billion: why the poorest countries are failing and what can be done about it*. Oxford University Press.

[16] BBC (2007) 'Billions face climate change risk', BBC News On-line, 6 April 2007, http://news.bbc.co.uk/1/hi/sci/tech/6532323.stm [accessed 8 July 2008].

[17] Stop Climate Chaos is a coalition of civil society organisations in the UK seeking to mobilise public concern to 'stop climate chaos'. See www.stopclimatechaos.org [accessed 8 July 2008]. Make Poverty History is an international civil movement seeking political change to alleviate absolute poverty around the world. See www.makepovertyhistory.org [accessed 8 July 2008].

This is, in essence, the question we first posed in Chapter 4: *The Endowment of Value,* when we examined the economics of climate change. There the answer was partly revealed in the value of the equity-weighting parameter used in cost–benefit analysis. This question also relates to those explored in Chapter 5: *The Things We Believe* about just solutions to climate change. It is no surprise to find that different answers to the question exist both in theory and as revealed by our actions. And the fact that there are different answers to this question is one of the reasons why the 2004 Copenhagen Consensus – *Global Crises, Global Solutions* – provoked such a storm of controversy. The Copenhagen Consensus, chaired by Bjørn Lomborg, drew attention to the economic rationale for investment in securing conventional development objectives rather than in mitigating climate change. It argued that, given $50 billion, it would be better to invest this amount in health care and poverty alleviation rather than in mitigating climate change. Preventing the deaths and poverty of today is a greater priority than reducing the climate risks of tomorrow.

It is, of course, easy to say that we should be seeking both to stop climate chaos *and* to make poverty history: the illusion that there is a choice to be made is just that – an illusion. Thus development economist Jeffrey Sachs argues, in his response to the Copenhagen Consensus: 'With the funds already promised, and which can easily be delivered, by developed nations … the world would not have had to choose between addressing specific diseases or overall health systems, or between [investing in] small-scale water projects and [mitigating] long-term climate change. We could do both.' [18] Indeed, by establishing eight Millennium Development Goals to be secured by 2015 *and* by negotiating the Kyoto Protocol, which sets emissions reductions targets for developed nations by 2012, the United Nations can claim to be forcing both agendas through their offices.

[18] p. 725 in Sachs, J. D. (2004) Seeking a global solution. *Nature* 430, 725–6.

Yet such cursory and easy claims-making does not help us understand why we do nevertheless disagree about how best to tackle poverty in the light of climate change and, more importantly, why poverty remains endemic in our world. As commented by the journal *Nature* in 2007, on securing the Millennium Development Goals: 'The eight goals themselves are ambitious to say the least. It will be a tall order to halve extreme poverty, roll back child mortality and provide universal primary education, all by 2015.' [19]

This brings us back to the different ways in which we think about the idea of sustainable development. If sustainability can be delivered through combinations of market environmentalism and ecological modernisation, modest adjustments to the prevailing development paradigm offer the prospect of adapting to climate change tomorrow by reducing poverty today. In this (modestly) adjusted model of traditional development, economic growth – even if it is 'green' growth – offers the most direct way to alleviate poverty today and thereby, indirectly through enhancing human capital, contributing to reduced climate vulnerability tomorrow. This is the development model being pursued, for example, by China (see Box 8.2).

Opponents of market environmentalism and ecological modernisation dismiss such development pathways as perpetuating the root causes of climate change. The fixation on growth, the assumption that increasing consumption is the way to well-being, is exactly the reason why greenhouse gas emissions continue to rise. This challenge to the dominant development paradigm is dismissed by others, for example Claire Fox from the London-based Institute of Ideas: 'It is not our excessive carbon emissions that deprive 1 billion people of clean water or doom the Earth's poorest to dependence on subsistence farming. Rather, it is ... man-made [sic] green fatalism that dismisses any possibility of development because "costs ... would be prohibitively

---

[19] Editorial (2007) The war on want. *Nature* 449, 947.

too high".'[20] People are vulnerable to climate change because they are poor; they are not poor because of climate change.

On the other hand, if the planet's resource limits – including its offering to humanity of relatively benign and accustomed climates – *are* being approached, an entirely different development paradigm may be needed. If the structures of a globalising and capitalising economic order are the reasons for this impending breakdown, more radical pathways to securing sustainability are needed. Eco-anarchists and eco-socialists rail against the 'economic domination of Northern interests and transnational corporations' because they lie behind both ills: climate chaos *and* the perpetuation of poverty. Indian environmentalist Sunita Narain captures this critique succinctly: 'All technofixes [for climate change] – biofuels, GM crops or nuclear power – will create the next generation of crisis, because they ignore the fundamental problems of capitalism as a system that ignores injustice and promotes inequity.'[21]

If justice becomes the organising principle around which responses to climate change are organised, then neither mitigating climate change nor adapting to its future effects are sufficient on their own. Something more is needed. In the eyes of some, such as Saleem ul Huq from the International Institute for Environment and Development in London, it is redress for past injustices. 'A major challenge now is to find ways to compensate people for the [climate] damage that has already been done.'[22]

Science again offers little here either as a moral compass or as an impartial adjudicator of such rival policy options: market environmentalism or ecological radicalism. The conclusions of the latest

[20] p. 5 in Hillman, M. and Fox, C. (2007) Carbon rationing: a valuable way of cutting carbon emissions. *Science and Public Affairs*, June.

[21] Narain, S. (2008) Editorial: Climate change: new opportunity to renew old orders. *Down to Earth* 15 (April). See www.downtoearth.org.in/default20080415. htm [accessed 10 July 2008].

[22] Huq, S. and Toulmin, C. (2006) *Three eras of climate change*. International Institute for Environment and Development: London.

IPCC reports may 'provide yet another reason to seek development pathways out of the pitiless poverty in which far too much of humanity is trapped',[23] yet they offer no pronouncement on whether those pathways towards sustainability should be reformist or radical.

8.4 Climate Change and Population

If arguments about the intersection of climate change and poverty are heated, how much more so are arguments about another dimension of sustainability which cannot be ignored – the relationship between climate change and population.

In the greenhouse gas emissions scenarios published by the IPCC in 2000, which continue to be widely used in scientific and policy analysis, three world population projections for the year 2050 were used: 8.7, 9.3 and 11.3 billion.[24] The difference between these three projections – 2.6 billion – represents more people than would be added to the world's population between today and 2050 – about 2 billion – under the lowest projection. The IPCC scenarios attracted considerable debate about the economic assumptions embedded in the scenarios, but remarkably little discussion has emerged about the significance for climate change of the optional 2.6 billion people by 2050. On the one hand, these optional 2.6 billion are people who will be emitting greenhouse gases in the future and contributing to climate change. Most of them will live in the burgeoning urban areas of developing countries. On the other hand, these optional 2.6 billion are also people who may emerge as victims of future climate risks exacerbated by climate change. Either way, whether the world's

---

[23] Editorial (2007) Climate change here and now. *Nature* 446, 701.

[24] IPCC (2000) *Emissions scenarios. Special Report of Working Group III of the Intergovernmental Panel on Climate Change.* Cambridge University Press. The equivalent population range for the year 2100 was from 7.1 to 15.1 billion. Today's (2009) world population is about 6.8 billion.

population in 2050 is closer to 8.7 or 11.3 billion is not predetermined; it can be influenced by decisions made today.[25]

It is very rare to find explicit discussion about population policies as part of the many debates about climate policy occurring around the world. Similarly, while there are numerous scientific analyses and policy debates about the desirable carbon dioxide concentration target (400, 450 or 550 ppm) or global temperature goal (e.g. 2° or 3°C above pre-industrial), there is very little discussion about a desirable global population or about preferred fertility rates. In their recent popular book on global warming, Gabrielle Walker and Sir David King mention population just once,[26] and in a context completely divorced from any policy discussion. As Chris Rapley, Director of London's Science Museum, has commented: 'So controversial is the subject [of population] that it has become the 'Cinderella' of the great sustainability debate – rarely visible in public, or even in private. In interdisciplinary meetings addressing how the planet functions as an integrated whole, demographers and population specialists are usually notable by their absence.'[27]

The Chinese government have made the most explicit and public statements about the relationship between climate policies and population policies (see Box 8.2). By adopting a 'one-child' policy since 1979, Chinese demographers estimate that about 300 million births have been avoided, equivalent to the present population of the USA. Even at the relatively low level of Chinese per capita carbon dioxide

[25] The most recent population estimates for 2050 from the UN (published 2006) suggest a world population between 7.8 and 10.8 billion (somewhat lower than that used by the IPCC), with a median figure of 9.2 billion. The range between the high and the low projections is now 3 billion people.

[26] Their sole mention of population was in this sentence: 'Even without the problem of climate change we would be finding our resources running thin. With it, the population boom becomes even graver.' p. 238 in Walker, G. and King, D. (2008) *The hot topic: how to tackle global warming and still keep the lights on*. Bloomsbury: London.

[27] Rapley, C. (2006) 'Earth is too crowded for Utopia'. BBC News On-line, 6 January 2006. See http://news.bbc.co.uk/1/hi/sci/tech/4584572.stm [accessed 14 November 2008].

emissions, the effect of this population policy can be measured as an avoidance of about 1.2 billion tonnes of carbon dioxide being emitted annually to the global atmosphere. This represents a nominal reduction of about 5 per cent in global carbon emissions, a much greater reduction than has been achieved by all the measures of the Kyoto Protocol.

Box 8.2: China's Climate Change (Development) Strategy

On 4 June 2007, China officially published its first national climate change strategy, *China's National Climate Change Programme* (CNCCP).[28] Led by the National Development and Reform Commission and linked to the country's eleventh Five-Year Plan, seventeen Chinese government departments contributed to the report, which was approved by the country's highest political body, the State Council.

The National Programme made clear that China regards climate change as much an issue of development as it is an issue of the environment. The priority in the country remains economic development, and with an economic growth rate averaging 9.5 per cent since 1979 its carbon dioxide emissions have quadrupled during this period. Yet China's cumulative historical emissions, and her per capita emissions rate, both remain low relative to developed nations. The National Programme makes clear that no international obligations to reduce emissions that challenge China's right to development would be acceptable.

Yet how does China plan to square the drive for economic growth, the delivery of development and the associated demand for energy,

---

[28] National Development and Reform Commission (2007) *China's national climate change programme*. Beijing, China.

with environmental protection and international environmental diplomacy? Three core elements are highlighted in the National Programme. First is the adoption of a voluntary target to reduce energy intensity in the country by 20 per cent over the period 2006–10, i.e. to drive forward energy efficiency improvements in the economy. Second is the renewed commitment to China's long-term afforestation and forest management programme which the Programme claims to have absorbed 5 billion tons of carbon dioxide since 1980. Forest cover is planned to increase from 12 per cent in 1980 to 26 per cent by 2050. A third contribution offered by China is to 'cash-in' the demographic benefits of her population policy of 'one child per family' and claim it as a climate mitigation policy. By avoiding an estimated 300 million births since 1979, the National Programme claims credit for avoiding annually about 1.2 billion tons of carbon dioxide emissions.

These claimed carbon credits reveal the difficulty of meeting the goals of development in a country with 1.3 billion people while containing the country's impact on global climate. The three emissions reduction credits are all 'nominal' in one sense or another: they represent emissions avoided relative to some hypothetical future scenario. With a burgeoning economy, rapid expansion of coal-generated energy and a cement-fuelled construction boom, China's absolute emissions of carbon dioxide into the atmosphere continue to rise at more than 5 per cent per year – and look like doing so for some time to come.

Why is there such reticence about debating global population alongside global temperature, about considering whether population policies might also be cost-effective climate policies? One of the reasons has been the political friction that emerged from the UN International Conference on Population and Development held in

Cairo in September 1994. The need to slow population growth in developing countries was noted, but the political emphasis of the Cairo Action Plan was on 'holistic approaches to reproductive health'.[29] The Conference was characterised by disputes about the ethics of China's 'one-child' policy, about the religious implications of promoting contraception and/or abortion, and about the coercive nature of some family planning proposals. One consequence of this fractious Cairo meeting is that population policies have slipped quietly down the international agenda; by 2004, investment by developed countries in international family planning programmes had fallen to just 13 per cent of the target set by the 1994 UN Conference.[30]

The intersections between climate change and population are clear. At its simplest level, the impact of humanity on the environment – including on global climate – can be viewed as a function of population, affluence and technology, the so-called 'IPAT formula'.[31] It would seem legitimate, therefore, to target all three components of this model if we seek to reduce human impact on the climate: containing population, limiting affluence, and cleaning technology. Indeed, US politician Al Gore recognised the legitimacy of this approach to climate mitigation way back in 1992, two years before the Cairo Population and Development Conference. 'No goal is more crucial to healing the global environment than stabilising human population.'[32]

Population should also be considered alongside climate change when considering the achievability of broad and universal development

---

[29] Campbell, M., Cleland, J., Ezeh, A. and Prata, N. (2007) Return of the population growth factor. *Science* 315, 1501–2.

[30] *Ibid.*, p. 1501.

[31] The IPAT formula was developed in the early 1970s by Barry Commoner and Paul Ehrlich within the framework of ecology as a way to think about and quantify the impact of humanity on the environment. The simplest form of the equation states that Impact = Population × Affluence × Technology (IPAT). A more recent variant of it is also known as the Kaya Identity, after Japanese economist Yoichi Kaya.

[32] p. 380 in Gore, A. (1992) On stabilising world population. *Population and Development Review* 18(2), 380–3.

goals. We commented earlier about the potential for climate change to delay or undermine securing the Millennium Development Goals by 2015, but similar arguments have been made that continued neglect of family planning in developing countries will also undermine the achievement of these goals. As reported by the UK All Party Parliamentary Group on Population, Development and Reproductive Health in January 2007: 'The Millennium Development Goals are difficult or impossible to achieve with current levels of population growth in the least developed countries and regions.'[33]

The thrust of these arguments leads to the conclusion that in the pursuit of both climate change mitigation and adaptation goals, containing and limiting the continued growth of the world's population makes sense. Demographer Brian O'Neill and colleagues at the International Institute for Applied Systems Analysis in Austria reached this conclusion at the end of their comprehensive survey of the relationships between population and climate change. 'Population-related policies ... reduce greenhouse gas emissions in the long run and improve the resilience of vulnerable populations to climate change impacts ... they easily qualify as no-regrets policies of the sort identified for priority action by the IPCC.'[34]

Yet, as we have seen, there has been very little connection between these two policy realms. Tackling climate change by tackling population growth is not a popular or politically correct approach in many quarters, as noted by American population ecologist Frederick Meyerson. 'Conservatives are often against sex education, contraception and abortion and they like growth – both in population and in the economy. Liberals usually support individual human rights above

[33] All Party Parliamentary Group on Population, Development and Reproductive Health (2007) *Return of the population growth factor: its impact on the Millennium Development Goals.* HMSO: London. See www.appg-popdevrh.org.uk [accessed 10 July 2008].

[34] p. 204 in O'Neill, B. C., MacKellar, F. L. and Lutz, W. (2001) *Population and climate change.* Cambridge University Press.

all else and fear the coercion label and therefore avoid discussion of population policy and stabilisation. The combination is a tragic stalemate that leads to more population growth.'[35]

It is argued by some that focusing on population policies detracts from the root causes of both poverty (e.g. unfair trade) and climate change (e.g. excessive consumption by elites). It is argued by others that population policies are in danger of impinging upon basic human rights, especially women's right to choose. China's coercive demographic regime, whatever its carbon emissions reduction benefits, does not sit easily with many religious, cultural and political sensibilities around the world. And yet one can't deny that the lifestyle decision that has the biggest single impact on a person's carbon footprint – especially a person in the developed world – is how many children to have. As estimated by statistician Paul Murtaugh, an average American woman deciding to avoid having a child can reduce her carbon legacy by about 800 metric tons of carbon dioxide, thereby reducing her own cumulative carbon footprint by almost 50 per cent without any other lifestyle changes.[36]

In the end, this debate about climate change and population reveals many of the fissures and divides that we have already identified in earlier chapters. Is climate change primarily about excessive consumption in the North or growing population in the South; about the alleviation of poverty today or about the mitigation of climate change tomorrow; about a revolution in energy technology or about a revolution in individual lifestyles? The different answers offered to these questions flow from the beliefs and values held by individuals and from their selective judgements about risk and precaution.

[35] Hartmann, B., Meyerson, F. A. B., Guillebaud, J., Chamie, J. and Desvaux, M. (2008) Population and climate change. *Bulletin of Atomic Scientists*, 16 April 2008, see www.thebulletin.org/web-edition/roundtables/population-and-climate-change [accessed 30 May 2008].
[36] Murtaugh, P. (2009) Reproduction and the carbon legacy of individuals. *Global Environmental Change* (in press).

While poverty (Section 8.3) and population (Section 8.4) both act as attractors for vigorous debates about the goals of sustainable development, they are not the only reasons why the challenges of development lead us to disagree about climate change. In the final section of this chapter we examine a number of other areas of potential conflict over climate change policies that are linked to sustainable development discourses.

8.5 Conflicts between Climate Change and Development

Conflicts between potential climate change policies and the goals of development and poverty alleviation arise in various arenas, offering fertile ground for disagreements about what to do about climate change. We examine here three such conflicts: the implementation of the Clean Development Mechanism; the desire to restrict the mobility of people and food; and the desire to displace fossil carbon in the transport liquid fuel mix through an expansion in global biofuel production.

Clean development. Under the terms of the Kyoto Protocol negotiated in 1997, developed countries which have accepted legally binding commitments to reduce greenhouse gas emissions are entitled to gain emissions 'credits' from investment projects that reduce emissions in developing countries. This mechanism is known as the Clean Development Mechanism (CDM) and links the policy fields of climate change and sustainable development. The twin goals of CDM projects are that they should assist developed countries to reach their emissions reductions targets under the Kyoto Protocol and also assist developing countries in achieving sustainable development. It is not at all clear that both of these goals are being simultaneously achieved.

Over 1,000 CDM projects have been financed since 1997, a large majority of these located in the middle-income economies of China,

India and Latin America. Studies which have examined the benefits of samples of these projects have concluded that there is often a trade-off between the goal of supplying cheap emissions credits and the promotion of sustainable development. In nearly all cases emissions credits have taken precedence in this trade-off.[37] Furthermore, the idea that the twin goals of emissions reductions and poverty alleviation can be achieved through traditional Overseas Development Assistance has also been shown to be flawed. German development expert Axel Michaelowa surveyed a large number of such projects and found that there were multiple conflicts between the objectives of climate mitigation and development policies. 'Most climate-change-related [development] assistance flows into medium-income emerging economies, and only addresses poverty alleviation indirectly, if at all. The CDM as a market mechanism will not contribute substantially to poverty alleviation either.'[38]

In such project investments, the promotion of developed-nation exports or an extension of political and cultural dominance is frequently given priority over the local development needs of the recipient communities. And it is very rare that CDM-driven investments into carbon offset projects end up yielding benefits for the poorest communities. On the contrary, the benefits usually accrue to urban middle-class elites in emerging economies. While many have argued that addressing the root causes of climate change using market-based mechanisms offers great opportunities to address simultaneously the issues of poverty, equity and sustainable development, performance to date has been disappointing.

Trade and mobility. In an increasingly globalised market with expanding trade in manufactured goods and food products across

[37] Olsen, K. H. (2007) The clean development mechanism's contribution to sustainable development: a review of the literature. *Climatic Change* 84(1), 59–73.

[38] p. 19 in Michaelowa, A. and Michaelowa, K. (2007) Climate or development: is ODA diverted from its original purpose? *Climatic Change* 84(1), 5–21.

national borders, the growth in international aviation seems inexorable. Nearly 25 per cent of China's carbon emissions, for example, arise from the manufacture of internationally traded goods which are transported around the world by plane and ship. Added to the growth of international tourism and commercial air travel, the 5 per cent per annum rise in greenhouse gas emissions from the aviation sector is increasingly the target of some climate activists and policy advocates. Reducing airport expansion, rationing flights, and 'food-miles' carbon accounting are policy responses emanating from more radical stances on sustainability that run counter to the prevailing economic paradigm of increased global trade and mobility.

Yet conventional development approaches, including those of market environmentalists, would argue that this liberalisation and expansion of trade and the growth in international tourism offers the best route out of poverty for many of the world's developing nations. International tourism provides the primary source of foreign exchange earnings in nearly fifty of the world's least developed nations. Shrink global tourism and, while aviation emissions may be contained, the economic sustainability of some of the poorest countries in the world will also be threatened. Replace conventional trade barriers in developed nations with discriminatory quotas based on the embodied carbon of imported goods, and developing-nation economies may again suffer. As summarised by British sustainability analysts Terry Dawson and Simon Allen: 'Attempts to control greenhouse-gas emissions may restrict international trade and tourism, removing the key strategies for less developed countries to grow their way out of poverty. A major disjunction is therefore looming between climate policy and strategies for reducing poverty.'[39]

Biofuels and food. Nowhere have the potentially contradictory goals of climate mitigation and sustainable development been more

[39] Dawson, T. P. and Allen, S. J. (2007) Poverty reduction must not exacerbate climate change. *Nature* 446, 372.

clearly exposed than in the case of the dash for biofuels which has characterised EU climate policy during the 2000s. The EU Directive on the Promotion and Use of Biofuels and Other Renewable Fuels for Transport came into force in October 2001. This Biofuels Directive stipulated that national measures must be taken by countries across the EU aimed at replacing 5.75 per cent of all transport petrol and diesel with biofuels by 2010, rising to 10 per cent by 2020.

The Directive was introduced as part of the EU's drive to reduce carbon dioxide emissions well below the 8 per cent reduction target stipulated in the Kyoto Protocol for the period 2008–12. But environmental and social concerns about the unanticipated side-effects of the Directive subsequently forced a re-evaluation by the Commission. New analysis showed that the net greenhouse gas savings from biofuels produced from rapeseed, corn (maize), sugar cane or palm oil were much smaller than originally thought and, in some cases, may actually lead to net increases in emissions relative to petrol and diesel sourced from fossil fuels. But it has been the social consequences of this policy, especially the impacts on the poorest communities around the world and their quest for food security, that have highlighted the conflict between sustainable development and this particular approach to climate mitigation.

The global sourcing of biofuels to meet the goals of the EU Directive contributed to land being switched from food crops to fuel crops and added a further incentive to convert tropical forest to soya and palm-oil production. As David Baldock of the Institute for European Environmental Policy commented back in 2003, 'We could end up importing vast quantities of soya and palm oil grown on newly deforested tropical land. It is a nightmare scenario, but not unimaginable.'[40] And the surge in global food prices during 2008 was linked in part to the EU's incentivising of biofuel production in pursuit of climate mitigation goals. The UN's special rapporteur on the right to food, Jean Ziegler, condemned the growing of biofuels as 'a crime against

---

[40] p. 9 in ECF (2003) *The biofuels directive: potential for climate protection? Conference Summary.* European Climate Forum: Tyndall Centre, UEA, Norwich, UK.

humanity' because they diverted arable land to the production of crops which are then burned for fuel instead of sold for food.[41] The outcome of such policy development is perverse. A policy designed in one part of the world to reduce the risks of future climate change threatening the viability of agriculture in other parts of the world has contributed to the short-term destabilising of the global food system. Rather than contributing to the theoretical food security of hundreds of millions of the world's poor in fifty years time, it has reduced the actual food security of tens of millions of the world's poor today.

The example of biofuels is used to illuminate one of the themes of this chapter; that our responses to climate change – through our understanding both of causes and of potential 'solutions' – are bound up with our pursuit of sustainable development. What may seem to be an attractive and viable 'solution' to climate change in Brussels – expanding the production of biofuels to displace fossil carbon from the liquid fuel mix – may undermine food security and poverty alleviation in, say, Burkina Faso. Such policy conflicts are not of course inevitable – food and fuel need not always be competing uses of land – but the different ways in which we weigh the importance of climate mitigation versus the goals of development open up many opportunities for discord in our policy advocacy.

8.6 Summary

So why do we disagree about climate change? We have shown in this chapter that the stories we tell about climate change and those we tell about sustainable development are inextricably linked. Even if sustainability is a destination we might broadly agree about in the abstract, the pathways of actually getting there – our conceptions and

---

[41] Quoted in Ferrett, G. (2007) Biofuels 'crime against humanity', BBC News On-line, 27 October 2007. See http://news.bbc.co.uk/1/hi/world/americas/7065061. stm [accessed 10 July 2008].

implementations of sustainable development – are many and varied. What this chapter has shown is that the relative significance of a billion people living in absolute poverty, an expanding global population, and the structural biases in a world trading system legated to us from the past is interpreted very differently by different people. For some, these characteristics of an unequal, unfair and unsustainable world remain serious obstacles to negotiating on climate change; they must be resolved first. For others, climate change ultimately trumps and overrides all of these challenges to sustainability and justice and is the priority to tackle.

The problem here is that these differences of perspective about climate change and sustainable development may imply that we are engaged in a zero-sum game where trade-offs have to be made. For some, attacking global poverty and reducing the vulnerability of billions to existing climate risks implies that the longer-term restructuring of the world's carbon-based energy systems gets neglected. For others, investing political, social and financial capital in family planning and reducing fertility rates in some parts of the world implies that a blind eye is being turned to excessive affluence in others.

As this chapter has shown, the way in which certain climate and development policies and measures have been implemented would seem to imply that trade-offs rather than synergies can indeed easily proliferate. But they need not. As American economist Jeffrey Sachs pointed out in his 2007 BBC Reith Lecture, there are some 'amazing bargains' to be found. 'My calculation has shown you that one day's Pentagon spending could cover every sleeping site in Africa for five years with anti-malaria bed nets. And yet we have not found our way to that bargain, the most amazing one of our time. We do have choices – they are good ones if we take them.'[42] Sachs' plea has, of course, been the

---

[42] Sachs, J. (2007) Lecture 1: 'Bursting at the seams', BBC Reith Lectures, Radio 4, broadcast 28 February 2007. See http://www.bbc.co.uk/radio4/reith2007/lecture1. shtml [accessed 14 November 2008].

petition of idealists for many years and many generations past, and it is one that is grounded in utopian visions, beliefs and priorities. But if this is what our examination of climate change and sustainable development brings us to, it also takes us forward to the future. Through our politics can we find sufficient grounds for agreement about climate change so that Sachs' 'amazing bargains' can be purchased?

All of the reasons for disagreement about climate change that we have so far examined find their voice when we articulate, dispute and negotiate the ways in which we wish to live together in society – which is, in the end, an exercise in politics. If politics is the ultimate social process we have designed for groups of people to reach, legitimate and enforce collective decisions, what are the prospects for reaching a political agreement about climate? Can we agree about how the world's climates should be governed? After all, we have considerable experience of using politics at local, national and even international scales to navigate our way through contentious issues and, at least in some circumstances, to secure stable and enforceable agreements. Whether such political experience can help us in the governing of climate change is what we will examine in the next chapter – *The Way We Govern*.

FURTHER READING FOR CHAPTER 8

Adams, W.M. (2008) **Green development: environment and sustainability in the South** (3rd edn). Routledge: London.
Adams offers a valuable analysis of the theory and practice of sustainable development. The book includes a survey of the roots of sustainable development thinking and its evolution in the last three decades of the twentieth century, as well as summarising several dominant and alternative discourses of sustainable development and sustainability. Examples are drawn from Africa, Latin America and Asia.

Munasinghe, M. and Swart, R. (2005) **Primer on climate change and sustainable development: facts, policy analysis and applications.** Cambridge University Press.
This book offers an introductory examination of the potential impacts and mitigation of climate change in the context of sustainable development. Future scenarios of development and climate change are presented, and an array of possible adaptation and mitigation

measures are discussed. Less committed and passionate than Porritt or Sachs, the book nevertheless provides a helpful road-map to the issues.

O'Neill, B. C., MacKellar, F. L. and Lutz, W. (2001) **Population and climate change.** Cambridge University Press.
This book is the best exploration yet published of the relationship between climate change and population. It examines the role of demographic factors in projections of future greenhouse gas emissions and also the impacts of climate change on population vulnerabilities. The editors argue that climate change strengthens the case for population policies.

Roberts, J.T. and Parks, B.C. (2007) **A climate of injustice: global inequality, North–South politics and climate policy.** MIT Press: Cambridge, MA.
This book adopts a strong normative position with regard to the politics of climate change. The authors argue that the current policy gridlock will remain unresolved until it is recognised that reaching a North–South global climate pact requires addressing the larger issue of inequality and striking a global bargain on environment and development. The book includes an analysis of the role that inequality between rich and poor nations plays in the negotiation of global climate agreements.

Sachs, J.D. (2005) **The end of poverty: how we can make it happen in our time.** Penguin Books: London.
Sachs draws on twenty-five years' experience to offer his vision of the keys to economic success in the world today. Combining anecdote and analysis, he draws a conceptual map of the world economy and explains why wealth and poverty have diverged and evolved across the planet, and why the poorest nations have been so powerless to escape the trap of poverty.

NINE

The Way We Govern

9.1 Introduction

On Tuesday 17 April 2007, the United Nations Security Council held a day-long debate on the security implications of climate change; a debate convened by the UK Government, which held the presidential chair for the month of April. It was the first time this particular United Nations body had deliberated upon climate change, and the discussion focused on the impacts of climate change on potential drivers of conflict, such as population movements, border disputes and access to energy, water, food and other scarce resources.

The debate featured interventions from more than fifty national delegations, representing imperilled island nations and industrialised greenhouse gas emitters alike. The session was chaired by the British Foreign Secretary, Margaret Beckett, and was the culmination of two years of international diplomacy by the UK Government seeking to re-frame climate change as a matter of global security. A year earlier, at a state function in Berlin, Beckett had laid out the agenda. 'Climate change is a serious threat to international security ... so achieving climate security must be at the core of foreign policy', and urged Europe to 'make climate security one of the continent's greatest priorities'.[1]

By bringing this debate to the Security Council, the UK Government was making a statement not just about the way climate change should be framed, but also about the way in which climate change should be governed. The United Nations Security Council is the organ of the UN charged with the maintenance of international peace and security. Its powers, outlined in the 1945 UN Charter, include the establishment of peacekeeping operations, the establishment of international sanctions regimes, and the authorisation for military action. It is the highest forum for deliberation among the nations of the world and comprises five permanently appointed national governments – the USA, Russia, China, the UK and France – and ten governments elected for two-year terms. It is the elite of the elite with respect to the negotiating power of sovereign governments.

Beckett justified this intervention on the grounds that climate change was a security issue not in the narrow sense of national security, but in the sense that it was about 'our collective security in a fragile and increasingly interdependent world'. She claimed the Council was not seeking to pre-empt the authority of other bodies such as the UN General Assembly, the UN Economic and Social Council, or the International Court of Justice. Instead, she argued, the decisions reached about climate change in these other bodies, and any subsequent actions taken, required the fullest possible understanding of the issues involved. 'Climate change can bring us together, if we have the wisdom to prevent it from driving us apart,' she declared.[2]

Not all speakers in the debate approved of this approach. There were particular reservations from developing countries, which saw climate change as a development issue to be dealt with by the more

---

[1] Speech by British Foreign Secretary Margaret Beckett, 24 October 2006, Berlin.
[2] Record of the 5663rd meeting of the UN Security Council in New York on 17 April 2007.

widely representative General Assembly of the United Nations. China's representative was among those who argued that the Security Council was not the proper forum for a debate on climate change, saying: 'The developing countries believe that the Council has neither the professional competence in handling climate change – nor is it the right decision-making place for extensive participation leading up to widely acceptable proposals.'[3] The issue could have certain security implications, but it was in essence an issue of sustainable development.

The representative of Pakistan – speaking on behalf of the 'Group of 77', a political coalition comprising developing countries and China – agreed, saying that the Council's primary duty was to maintain international peace and security. Other issues, including those related to economic and social development, should be assigned to the Economic and Social Council and to the General Assembly. The ever-increasing encroachment of the Security Council on the roles and responsibilities of the other main organs of the United Nations represented a 'distortion' of the principles and purposes of the Charter, infringed on the authority of the other bodies, and compromised the rights of the Organisation's wider membership.

But Papua New Guinea's representative – Robert Aisi – who spoke on behalf of the Pacific Islands Forum, said that the impact of climate change on small islands was no less threatening than the dangers posed to large nations by guns and bombs. Pacific island countries were likely to face massive dislocations of people, similar to large-scale population movements sparked by armed conflict. The impacts on identity and social cohesion were likely to cause as much resentment, hatred and alienation as any refugee crisis. 'The Security Council, charged with protecting human rights and the integrity and security of States, is the paramount international forum available to us,' he said.[4] The Forum did not expect the Council to get involved in negotiations

---

[3] *Ibid.*   [4] *Ibid.*

under the UN Framework Convention on Climate Change, but it did expect the fifteen-member body to keep the issue of climate change under continuous review. It also expected the Security Council to ensure that all countries contributed to solving the problem of climate change and that those efforts were commensurate with their resources and capacities. The Pacific Islands Forum called upon the Council to review sensitive issues associated with climate change, such as implications for sovereignty and international legal rights from the loss of land, resources and people.

This UN Security Council debate about climate change as a security issue illuminates a problem at the very heart of political deliberations about climate change – 'Who governs climate?' Climate change has been seen as a legitimate public policy issue since the late 1980s; an issue about which we believe sovereign governments not only have a duty to comment, but also an obligation to lead and to legislate. Internationally, the Kyoto Protocol, negotiated in 1997, has been the benchmark agreement for shaping the goals of (and disputes around) climate mitigation policies, while the last few years have seen growing attention paid to whether or not new international climate adaptation policies are necessary or desirable.

Given that global climate is now widely seen to be in need of some form of governance, who exactly has the right to govern climate, who exactly should 'exercise authority over' the global climate system? Is it the role of the elite of the governing elite, the UN Security Council, or is it the role of the UN General Assembly? Is it through the institutions of national sovereign governments that climate can and should be governed? And how are other, non-governmental, interests represented in any climate governance regime: the interests of indigenous peoples, multinational corporations, religious and social movements, individual citizens, non-humans? This chapter explores the various ways in which we grapple with these questions of climate governance and shows that **one of the reasons we disagree about climate change is because we seek to govern in different ways**.

Disagreements emerge because of different ways of understanding and exercising power, because of different political ideologies that influence the design of policy, and because of differences about whether climate policy is approached from unilateral, bilateral or multilateral perspectives. Within the UN Framework Convention on Climate Change, the Kyoto Protocol has emerged as the primary outcome of multilateralism. The Protocol is examined in Section 9.2 and is shown to place a high value on the traditional role of the nation-state as the main actor in international diplomacy. The Kyoto Protocol is a top-down approach to climate governance, one that fits with the idea of 'green governmentality'.

The way in which climate change is framed – as an environmental, security, economic or social justice problem – fundamentally affects the way that climate is governed, the types of policy interventions that are sought, and the actors to be involved. Ideology also has a major bearing on the nature of the climate policies espoused, whether an ideology which draws upon 'market environmentalism' (thus favouring emissions trading and personal carbon allowances, for example) or one that seeks a more interventionist role for the state (thus favouring regulation and taxes). The emergence of the international carbon market as the *de facto* favoured instrument of climate governance is explored in Section 9.3.

A third emergent dimension of climate governance has been called 'civic environmentalism' (Section 9.4). Rejecting top-down governance through the state and challenging the dominance of the carbon market, civic environmentalism seeks to transcend political ideologies and build coalitions of multi-level policy actors beyond the state: businesses, municipalities, citizens' groups, international non-governmental organisations. As with the discourses around sustainable development that we introduced in Chapter 8: *The Challenges of Development*, civic environmentalism has both reformist and radical variants.

Finally, in Section 9.5, we return to the Kyoto Protocol and the apparent mismatch between the demands of climate governance and the traditional governing role of the nation-state. We examine two aspects of this mismatch: so-called 'clumsy governance', eclectic alternatives to the universal and top-down governance offered by the Kyoto Protocol, and the challenges of governing newly proposed geo-engineering schemes.

9.2 Governing Climate Through Kyoto

We have already seen in Chapter 1: *The Social Meanings of Climate* and Chapter 2: *The Discovery of Climate Change* how two quite radical ideas about our physical climate emerged in the late twentieth century. First, climate began to be described and explained as the out-working of a fully interconnected global physical system and, second, human modification of the global atmosphere was changing this system to yield new and warmer climates. International collaborative science, promoted through bodies such as the World Meteorological Organization and assessed by institutions such as the IPCC, was essential for the development and promulgation of these ideas. This scientific concept of a global climate system being modified (inadvertently) by human actions has, more recently, adopted the language of Earth system science and analysis.[5] It has also spawned a new historical era – the Anthropocene: the era when humans first began to have a marked effect on the Earth's climate and ecosystems.[6]

---

[5] For a classic account of this approach, see Schellnhuber, H.-J., Crutzen, P., Clark, W.C., Claussen, M. and Held, H. (eds) (2004) *Earth system analysis for sustainability*. MIT Press: Cambridge, MA.

[6] This term was coined in the year 2000 by German atmospheric chemist and Nobel Prize winner Paul Crutzen. He suggests that the Anthropocene started in the eighteenth century, although paleoclimatologist Bill Ruddiman has suggested it began as early as 8,000 years ago.

A corollary of such new ways of understanding the physical climates we experience, and of the novel appreciation of humanity's influence on these climates, has been that managing such a globally connected system requires new global forms of governance. If the atmosphere truly offers no boundaries to the circulation of greenhouse gases around the planet, then a commensurate global system of climate governance must also break down the national and sectarian barriers of traditional forms of governance. The pioneer of this form of thinking about global environmental governance was the Vienna Convention for the Protection of the Ozone Layer, agreed in 1985. This Convention led to the Montreal Protocol, which was signed in September 1987 by forty-six national governments (191 countries have subsequently ratified it) and which has operated as the framework through which ozone-depleting substances have been gradually eliminated from the global industrial economy.

It became the established wisdom for many that the Montreal Protocol offered a prototype for the sort of international treaty and governance regime that the growing concern about climate change seemed to demand (see Box 9.1 for a discussion about the merits and demerits of such an analogy). Self-consciously modelling itself on the Vienna Convention, the UN Framework Convention on Climate Change was signed in June 1992 at the Rio Earth Summit, followed by five years of negotiation which led to the Kyoto Protocol being agreed in December 1997. The Protocol required industrialised nations to reduce their collective greenhouse gas emissions by 5.2 per cent by the period 2008–12, relative to 1990 levels, although this reduction could be achieved through a variety of flexible trading and investment mechanisms. Considerable political prevarication, diplomacy and renegotiation then ensued, as countries felt the need to protect their national economic self-interest. It took a further seven years – until February 2005 – for enough signatories to be gathered and for the Protocol to enter into force.

**Box 9.1: The Montreal Protocol: A Prototype
for Climate Governance?**

The 1987 Montreal Protocol on Substances that Deplete the Ozone Layer is regarded by many as the most successful international environmental treaty. Even though it may not have had much competition for being awarded this status, it is not hard to see why this view has prevailed. The worldwide consumption of ozone-depleting substances fell by over 90 per cent between 1986 and 2004, and the depletion of stratospheric ozone over Antarctica has stabilised, with projections suggesting a recovery close to 1980 levels by the 2050s. It is not surprising, therefore, that the Montreal Protocol exerted a powerful influence on climate change negotiators in the early 1990s, as they started thinking about how global climate was to be governed. As climatologist Steve Schneider remarked in 1998 after the signing of the Kyoto Protocol: 'It took an ozone hole to finally persuade the major nations of the world to act to control ozone-depleting substances. While that example holds open the optimism that international action across diverse interests is possible, it also reminds us that demonstrable damage was needed to get our attention.'[7]

The 'optimism' derived from the success of Montreal, together with the generally unquestioned assumption that nation-states hold the key to governing the global atmosphere, led to the same binding targets-and-timetables approach being applied to greenhouse gas emissions reduction. Led by the European Union and by environmental policy advocates, the Kyoto Protocol set 'quantified limitation and reduction objectives within specified time frames' to be achieved by named sovereign states. There was little

---

[7] p. 18 in Schneider, S. H. (1998) Kyoto Protocol: the unfinished agenda. *Climatic Change* 39, 1–21.

consideration given, however, as to exactly *how* these reductions were to be delivered by the governments who signed the Protocol.

Yet the problem structure of global climate change is fundamentally different from the problem structure of stratospheric ozone depletion. Their superficial similarity – both being related to gaseous emissions into the atmosphere and having multi-decadal environmental consequences – is dwarfed by the many differences that exist in the causes, consequences and actors involved. The causes of ozone depletion were a limited number of artificial gases released through production and consumption practices that involved a tiny fraction of global economic activity. The causes of anthropogenic climate change emerge from nearly every aspect of human activity – the use of energy and the use of land – and from a very large proportion of global economic production. The consequences of ozone depletion were relatively limited and quantifiable. The consequences of climate change again affect nearly every aspect of human activity and remain largely intractable to our foresight. The actors involved in ozone depletion were primarily two large multinational chemical corporations and a relatively small number of countries which housed manufacturing plants. The actors involved in climate change are every person on the planet, most public- and private-sector organisations and, to varying levels of significance, all countries.

Climate change is also different from stratospheric ozone depletion in the way in which individuals and cultures relate to the two phenomena. As we saw in Chapter 1: *The Social Meanings of Climate*, humanity has a very long cultural history of understanding the idea of climate and of experiencing its sensual intimacy. There is no such cultural or sensual history of relating to stratospheric ozone. While both phenomena are largely invisible to our senses, revealed through the constructions of environmental science, climate and ozone occupy radically different sites and fulfil quite different roles in our cultural imaginations.

When these various differences are understood, there seems to be no *a priori* reason why what worked well for governing stratospheric ozone should also work well for governing global climate. The irony is that, while the Kyoto Protocol was modelled – perhaps poorly – on the Montreal Protocol, it is the Montreal Protocol which has to date offered the greater climate change mitigation benefits. As an indirect consequence of reducing ozone-depleting substances, which also act in the atmosphere as greenhouse gases, the Montreal Protocol has reduced the human effect on climate by a much greater amount than have the emissions controls stimulated by the Kyoto Protocol. As shown by Dutch atmospheric chemist Guus Velders and colleagues, 'the climate protection already achieved by the Montreal Protocol alone is far larger than the reduction target of the first commitment period of the Kyoto Protocol'.[8]

The Kyoto Protocol is a classic example of an international treaty between sovereign nation-states in which national targets and time-tables are set for controlling behaviour, in this case behaviour and activities that produce emissions of greenhouse gases. It adopts an approach to understanding governance arrangements that international relations experts call 'regime theory'. Nation-states, in the absence of a world government or supranational authority, seek to 'regulate their practises in a fairly well co-ordinated manner in order to solve collective action problems'.[9] The Protocol emphasises the

---

[8] p. 4814 in Velders, G. J. M., Anderson, S. O., Daniel, J. S., Fahey, D. W. and McFarland, M. (2007) The importance of the Montreal Protocol in protecting climate. *Proceedings of the National Academy of Sciences* 104(12), 4814–19.

[9] p. 5 in Okereke, C. and Bulkeley, H. (2007) *Conceptualizing climate change governance beyond the international regime: a review of four theoretical approaches.* Tyndall Centre Working Paper No. 112. University of East Anglia, Norwich, UK.

supremacy of the nation-state in governing, the power of the nation-state in delivering emissions reductions goals, and the legitimacy of the nation-state for mobilising and regulating its citizens. When placed in the context of the organisational and negotiating structures of the United Nations, this is the line of thinking that ends with climate change being debated in the UN Security Council.

The Kyoto Protocol has proved remarkably durable despite being at the centre of a continuing sequence of disputes and disagreements since it was first signed in December 1997. It was forged through five years of hard diplomacy between Rio and Kyoto. It survived seven years of uncertainty and prevarication between its signing and its eventual coming into force. It survived the 2001 withdrawal of the largest single emitter, the USA, as President George W. Bush described the Protocol as 'fatally flawed'. And it survived the attacks from critics who claimed that the emissions reduction targets agreed were ineffective in slowing climate change, that the mechanisms to achieve the nationally set targets were too liberal, that the negotiating structure was too complex, and that the sanctions for non-compliance were non-existent. As demonstrated by this news report from 2005, the Kyoto Protocol has refused to lie down. 'Kyoto Protocol confirmed as the only game in town: Critics damn it for a long list of reasons and it has been declared dead several times, but the Kyoto Protocol emerged stronger than ever after the Montreal conference on climate change that ended here Saturday.' [10] We examine later in the chapter alternative approaches to the governing of climate, but we first need to ask why it is that the Kyoto Protocol elicits such strong feelings from both its advocates and its critics.

On one side are those who applaud the combination of nation-state multilateralism and legally binding national emissions reductions

---

[10] Agence-France Presse (2005) 'Kyoto Protocol confirmed as the only game in town', Montreal, 12 December. See www.terradaily.com/reports/Kyoto_Protocol_Confirmed_As_The_Only_Game_In_Town.html [accessed 10 July 2008].

targets as essential cornerstones of any climate governance regime. Quantified emissions reduction targets, woven together across nations into a universal framework that unites all actors and serves as a locus for debate and decision making, lies at the heart of this conception of global climate governance. Eileen Claussen, President of the Pew Center on Global Climate Change in the USA, is clear about what is needed. Prior to the adoption of the Bali 'road-map' during the 13th Conference of the Parties (COP-13) to the Framework Convention in December 2007, Claussen argued that, 'the critical issue in Bali is whether the new road map sets governments on a course towards establishing binding multilateral commitments ... integrated into a common framework'.[11]

Such advocates of Kyoto point to the successful creation of market-oriented institutions and rules, including international emissions trading of six different greenhouse gases. Carbon emissions (and hence also avoided carbon emissions) have a market value and the marginal costs of reducing emissions are lowered through trading. Supporters of the Protocol also draw attention to aspects which play well to concerns for distributional justice. Thus binding commitments are for those countries which have borne the greatest historical responsibility for emissions and which, generally, have a greater ability to pay. This is consistent with the 'polluter pays' principle; a concept in environmental law where the polluting party pays for the environmental damage caused by their actions.

For others, however, such as American economist Scott Barrett, the Kyoto Protocol and its guiding philosophy has been an example of everything that an international environmental treaty should not be. Barrett describes it thus: 'To be successful, a treaty must fulfil three conditions ... it must attract broad participation ... it must ensure compliance ... and it must do both of these things even as it asks its

[11] Claussen, E. (2007) 'The road from Bali: towards a binding post-2012 climate framework'. See www.climateactionprogramme.org [accessed 30 May 2008].

parties to change their behaviour substantially. It is easy for a treaty to meet one or two of these conditions. It is very hard for a treaty to do all of them. Kyoto fails to do any of them.'[12]

The failures of the Kyoto Protocol are clear under such criteria. Three of the four largest greenhouse gas emitters – China, the USA and India[13] – are not participants in the Kyoto Protocol in the sense that they have not taken on any emissions reductions obligations. The Protocol has very weak compliance mechanisms; so weak, in fact, that nations can completely ignore their reduction obligations with impunity – as in the case of Canada, where emissions by 2010 are predicted to exceed their Kyoto target by 45 per cent[14] – or else use flexible mechanisms which allow nations to purchase 'reduction credits' elsewhere in the world. And Kyoto has hardly been successful in inducing parties to change their behaviour. The negotiation of flexible mechanisms – the inclusion of sinks, Russian 'hot air', the Clean Development Mechanism – means that few substantial domestic actions have been taken to induce absolute reductions in domestic greenhouse gas emissions. Even the UK, one of the strongest supporters of the Kyoto Protocol and despite a variety of climate change measures and policies, has seen national carbon emissions rise by about 3 per cent during the decade 1997–2006.[15]

The Kyoto Protocol falls into a category of environmental governance which Swedish political scientist Karen Bäckstrand calls 'green

---

[12] p. 2 in Barrett, S. (2008) *Climate change negotiations reconsidered*. Briefing Paper for Policy Network: London.

[13] The third largest emitter is Russia, and although they *have* ratified Kyoto, they benefit from the Protocol through the so-called 'hot air' (emissions credits) that the collapse of the Soviet Union introduced into their greenhouse gas emissions profile.

[14] Barrett, *Climate change negotiations reconsidered*, p. 3

[15] The rise in emissions over this period increases to well over 10 per cent if the contributions from the rapidly growing aviation and marine transport sectors are included, along with embodied carbon in imported goods.

governmentality'.[16] This green governmentality of climate has emerged alongside the science-driven framing of climate change as a global environmental problem. In this case, green governmentality adopts a centralised and bureaucratised multilateral negotiation system and places the nation-state at the centre of the governance arrangements. In terms of Cultural Theory (cf. Chapter 6: *The Things We Fear*) we can view the Kyoto Protocol as a form of global climate governance which sits very easily with a hierarchist way of life – order, control and multiple sets of rules governing social structures. But there are other forms of climate governance which adopt different ideological and political stances and which are more congenial to non-hierarchist ways of life. Kyoto is *not* the only game in town. We will explore two of them in Sections 9.3 and 9.4 – approaches to climate governance which we might associate with the terms 'market environmentalism' and 'civic environmentalism'.

9.3 Governing Climate Through the Market

One of the central pillars of the Kyoto Protocol has turned out to be the international trading of carbon emissions. Carbon trading has been modelled on the sulphur dioxide trading scheme developed in the USA during the 1980s. Using the power of the market to drive down costs, sulphur dioxide trading reduced the costs of emissions reductions in this country by a factor of between two and ten. By assigning property rights for greenhouse gas emissions and establishing a market for the trading of these rights, carbon trading now seeks to harness this same power of the market in the service of climate mitigation. Climate governance thus becomes less the responsibility of national legislatures through command-and-control measures and more the responsibility of markets and their traders.

---

[16] Bäckstrand, K. and Lövbrand, E. (2008) Climate governance beyond 2012: competing discourses of green governmentality, ecological modernisation and civic environmentalism, in Pettenger, M. (ed.), *The social construction of climate change*. Ashgate: Aldershot, UK, pp. 123–41.

Although international carbon trading has become almost synonymous with the Kyoto Protocol, the two are not necessarily dependent on each other. In the negotiations leading to the Kyoto Protocol, for example, both the European Union and many of the more determined environmental policy advocates were adamant that trading carbon emissions should not be part of their preferred targets-and-timetables approach. Domestic command-and-control approaches to reducing emissions were seen as preferable to relying on the market to deliver. Under the pressure of negotiation, however, especially from the USA, this position was relaxed and now the EU operates the world's largest carbon market valued in 2007 at €28 billion.

Conversely, international carbon trading does not depend on the existence of a universal treaty underpinned by national targets and timetables. The EU emissions trading scheme was put into operation *before* the Protocol entered into force and its first accounting period (2005–7) sat outside the requirements of the Protocol. And at least five other national (or limited international) carbon markets emerged around the world before the Kyoto Protocol came into force (see Figure 9.1).

Yet there is no doubt that the Kyoto Protocol has provided a catalyst for the development and experimentation of regional and global carbon markets. Carbon trading has emerged as the policy instrument of choice for governing climate. International carbon trading is set to continue to feature as the cornerstone in any post-2012 extension to the Kyoto Protocol and it finds enthusiastic backers both from business entrepreneurs and from politicians. Thus Tom Whitehouse of the UK-based consultancy, Carbon International, claims that 'carbon is a new commodity, a new currency. I believe a robust carbon market can and will deliver the emissions cuts that will delay and avert climate change';[17] while EU Environment Commissioner Stavros Dimas

---

[17] Quoted on p. 38 in Pearce, F. (2008) Dirty, sexy money. *New Scientist*, London, 19 April.

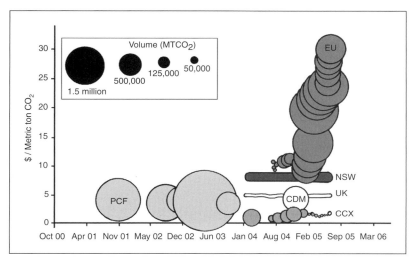

FIGURE 9.1: *Prices (vertical scale) and volumes (size of circles) for six different emissions trading schemes – Prototype Carbon Fund (PCF), the Clean Development Mechanism (CDM), the EU Emissions Trading Scheme (EU), the Chicago Carbon Exchange (CCX), the UK trading scheme (UK) and the New South Wales, Australia, trading scheme (NSW).*
Source: Victor et al. (2005).

is equally enthusiastic: 'The EU fully recognises the crucial value of such market-based mechanisms to harness the creativity of the business sector, to offer economic incentives to cut emissions, and to reduce compliance cost. The Emissions Trading Scheme will reduce the cost of achieving the EU's Kyoto target by about a third.'[18]

Carbon trading may also arise in new contexts, not necessarily mandated by the Kyoto Protocol. Thus there seems every prospect of a national cap-and-trade Bill soon being passed in the USA. Thirteen greenhouse gas emissions reduction Bills were put before Congress during 2007, a majority of them proposing some form of market-based carbon trading mechanism. The Senate led the way with the

[18] Speech by Stavros Dimas (2005) to the American Chamber of Commerce, 19 July, Brussels.

Lieberman–Warner Bill, also known as America's Climate Security Act. Carbon trading also finds its advocates at the level of individual citizens. The idea of tradable personal carbon allowances has been widely researched and debated in the UK and has already been the subject of one Bill presented before the British Parliament. By linking the power of the market with individual consumer choice, the argument is that personal carbon allowances offer a more empowering option for climate governance than do more directive forms of state coercion. This was certainly the argument of the British Secretary of State for the Environment, David Miliband, when he praised the concept in 2006: 'It is a compelling thought experiment – limit the carbon emissions by [individual] end users based on the science, and then use financial incentives to drive efficiency and innovation.' [19]

Carbon trading emerges favourably from an understanding of sustainable development which operates under the label 'market environmentalism' (cf. Chapter 8: *The Challenges of Development*). This neo-liberal approach to environmental governance and natural resource management – what has been termed somewhat pejoratively as 'commodifying the environment' – has become increasingly favoured in recent years, as noted by political geographer Ian Bailey: 'This trend ... has radically rewritten the priorities of environmental policy ... cost-efficiency, competitiveness and ... self-, or co-steering, market processes have been [favoured] over government mandates as the core elements of effective environmental regulation.' [20]

Governing climate through the ideology of market environmentalism lies behind two other emerging policy instruments for climate mitigation: payment for ecosystem services in the form of avoided tropical deforestation, and voluntary carbon offsets. As part of the preparatory

[19] Speech by David Miliband (2006) The great stink: towards an environmental contract, 19 July, London.
[20] p. 530 in Bailey, I. (2007) Market environmentalism, new environmental policy instruments and climate policy in the UK and Germany. *Annals of the Association of American Geographers* 97(3), 530–50.

work for a post-2012 extension of the Kyoto Protocol, consideration has been given to developing market-based approaches for incentivising tropical countries to reduce carbon emissions from deforestation, the so-called Reducing Emissions from Deforestation and Degradation proposal. Financial payments or tradable carbon credits would be offered by the international community, or by private investors (see the example of Canopy Capital cited in Chapter 8), to countries that ensured that their tropical forests continued to act as carbon sinks.

The other example is the rise of the voluntary carbon offset market, particularly in Europe and North America. Carbon offsets have emerged from the market logic of pricing and trading carbon emissions reductions (for example the Clean Development Mechanism) and offer a certified reduction in carbon dioxide emissions somewhere in the world in return for a payment (an 'investment'). In the case of voluntary offsets, these carbon 'credits' are purchased by individuals, organisations or small companies who are not participants in any of the formal emissions trading schemes (for example those shown in Figure 9.1). The first voluntary carbon offset deal was brokered in 1989 and the market has grown exponentially in the last few years. In 2007 there were over sixty different organisations in the UK alone offering such voluntary carbon offsets.

Given our earlier exploration of differences in beliefs, values and ways of life, it is not surprising that strident challenges to the principle of governing climate through the market have been raised. These are not just challenges to the practicalities of carbon trading, such as the design of allocation rules, emissions reduction monitoring and compliance, or the definition of 'additionality'.[21] They are challenges to the very idea that extending property rights to the atmosphere and

---

[21] 'Additionality' is the idea that emissions reductions arising from a deliberate carbon-offset investment must be over-and-above, i.e. additional to, any reductions in emissions that would have occurred without such an investment. Additionality is therefore a rather elusive property of carbon projects and can never be proved, only assumed on the basis of reasonable counterfactual scenarios.

then trading those rights through a market is the right way to manage the environment.

The critique of market environmentalism as it is applied to climate change operates at different levels. One reaction is that carbon trading – seeking emissions reduction at lowest cost with little consideration to geography – avoids the need to confront questions of justice and equity. These questions arise because of the historical responsibility for past greenhouse gas emissions and because of the contemporary responsibility to ensure equitable distribution of the benefits of carbon emissions reductions. This latter concern is especially acute, for example, amongst indigenous forest communities in developing countries. Investing in emissions reductions projects at lowest cost may be good for economic efficiency in the global economy, but such investments rarely demand that attention is paid to these dimensions of historical and social justice. Carbon trading, and the associated rise of carbon financiers, may just be the latest form of colonialism – a 'carbon colonialism', as coined by campaigner Larry Lohman, founder of the Durban Group for Climate Justice.

At a deeper level, though, is a broader critique of the commodification instinct, the instinct to extend the market into novel realms. Some react against this transference of power from state to market as wholly inappropriate for the scale of the challenge that they perceive the mitigation of climate change to demand. As recounted by science journalist Fred Pearce: 'Many fear the carbon capitalism is already out of control, delivering big profits while doing little to halt global warming. They are deeply sceptical of the notion that market forces can fix climate change. "To believe that is to believe in magic," says Tom Burke, a former director of Friends of the Earth.'[22]

This commodification of the atmosphere is resisted as part of a wider critique of contemporary governance: the retreat of government from the public sphere and the favouring of privatisation of what are

---

[22] Pearce, Dirty, sexy money, p. 38.

deemed to be public goods and services. UK political scientist Andrew Dobson challenges the assumption implicit in the commodification of carbon – that people will only act for the common good when it's in their personal interest to do so. Rather than building the governance of climate change around private contracts and commercial markets and the philosophy of 'I will if you will', which results in 'I won't if you won't', Dobson demands a different social logic for the governance of the largest of our public commons, the global atmosphere: 'I will even if you won't'. In the end, 'The biggest casualty of the rush to privatisation, enclosure and the withering of the public sphere may well be the climate itself.' [23]

The rise of international carbon trading – and the associated variants such as Reducing Emissions from Deforestation and Degradation and voluntary carbon offsetting – also offers a challenge to the pre-eminence of the nation-state in the negotiating and governing of climate change. International emissions trading schemes require national governments to accede to other supranational authorities, whether the EU in the case of the European Trading Scheme or the UN Framework Convention in the case of the Clean Development Mechanism. The private sector – investors, companies, traders – become the agents of climate governance and national governments become second-line managers merely to set and administer the rules. Similar challenges to state sovereignty and hegemony – challenges to the idea that 'regime theory' adequately describes the relations between nation-states – have been observed in other areas of international relations. The forces of globalisation are fracturing the traditional international governance model of a community of independent and sovereign nations. Ian Bailey observes it thus: 'Neoliberal climate policies at the international level [i.e. carbon trading] is intensifying

---

[23] Dobson, A. (2008) 'Climate change and the public sphere', OpenDemocracy, 1 April. See www.opendemocracy.net [accessed 10 July 2008].

and producing new patterns of interaction between supranational, national and non-state actors.'[24]

We next turn to consider other roles and partnerships for these new patterns and emerging non-state actors and whether they offer a way of governing global climate that may engender greater social and political agreement.

9.4 Climate Governance Through Non-State Actors

We have seen how the early moves in the 1990s to establish a form of global climate governance through the UN Framework Convention and the Kyoto Protocol was predicated on the agency of the sovereign nation-state. The idea of 'regime theory' underpins attempts to understand such an approach to governance in which the autonomy, control and authority of the nation-state is first assumed and then deployed. We have also surveyed a favoured instrument of this putative climate regime – the emerging carbon market, and its roots in the ideology of market environmentalism expressed most visibly through the new mode of carbon capitalism.

There is resistance to both of these dimensions – the 'who' and the 'how' – of climate governance. In the eyes of some, the idea of a single universal climate treaty, negotiated and agreed by 180 sovereign nations, is a licence for delay, obfuscation and, eventually, the emasculation of climate policies. For example, German political scientist Hermann Ott and colleagues have remarked on the daunting prospect of negotiating a new post-2012 extension of the Kyoto Protocol, following the tentative template reached in Bali at COP-13. 'All these processes have to be kept under surveillance, kept apart, streamlined where necessary, and – in the end – all those different threads have to be combined into one gigantic package deal. Tying all the pieces

---

[24] p. 440 in Bailey, I. (2007) Neoliberalism, climate governance and the scalar politics of EU emissions trading. *Area* 39(4), 431–42.

together has already proved difficult enough in Bali.'[25] Others resist the hegemony of the carbon market in climate governance and all that this implies for the relationship between Nature, capital, state and citizen. The nation-state is too limited in its freedom of manoeuvre, having to work through disputed and sluggish international treaties. Yet the carbon market, through its appropriation of powers to reshape climate according to the invisible 'green hand' of carbon capitalism, is operating without sufficient limits.

Out of such dissatisfaction emerges a different approach to climate governance, an approach which challenges both the state-centred beliefs of green governmentality (instinctively favoured by hierarchists) and the market-centred beliefs of neo-liberal environmentalists (instinctively favoured by individualists). It is an approach called 'civic environmentalism' and places citizenship, participation and equity at its core. It is egalitarians who might find more sympathy with this approach to climate governance.

The reform-oriented version of civic environmentalism 'highlights how the vital force of a transnational civil society, which serves as a complement to state-centric practises, can increase the public accountability and legitimacy of the climate regime'.[26] In this conception of agency and governance, it is non-state actors who take centre-stage and who hold the key to effecting reductions in carbon emissions. Businesses, trades unions, city authorities, women's groups, carbon offset companies – even the Courts and celebrities – become effective agents of change. Such non-state actors, of course, operate within the diplomatic framework of nation-states and are also participants in emerging new carbon markets, but they do not see themselves as dependent on either the state or the market to effect change. They see themselves as more effective, more efficient and more rapid at tackling

[25] p. 93 in Ott, H. E., Sterk, W. and Watanabe, R. (2008) The Bali roadmap: new horizons for global climate policy. *Climate Policy* 8(1), 91–5.

[26] Bäckstrand and Lövbrand, Climate governance beyond 2012, p. 124.

climate change than are nation-states, and they see themselves as more able to attend sensitively to concerns of fairness, democracy and participation than can the market. We will examine three examples of climate governance beyond the state: the Local Governments for Climate Protection Agreement, the Courts, and Carbon Reduction Action Groups.

Local government. Many local and municipal governments have developed their identities as entities responsible for incorporating concerns about climate change into their long-term strategies and policy making. The most visible international manifestation of this civic movement was the signing, during COP-13 in Bali in December 2007, of The World Mayors and Local Governments Climate Protection Agreement.[27] The Agreement committed the voluntary signatories to reduce greenhouse gas emissions 'immediately and significantly', eventually contributing to a worldwide goal of reducing emissions by 60 per cent relative to 1990 levels. This multilateral initiative operates at sub-national levels of governance.

Courts. Given the scientific claims of cause and effect between greenhouse gas emissions, climate change and the impacts of climate risks, it is not surprising to find that various civil actions have been brought before the Courts for legal adjudication. This has been particularly the case in the USA. Frustration with the lack of Federal legislation on greenhouse gas emissions has prompted individual citizens and some civic groups to act as plaintiffs against large corporations. The charge is that large volumes of emissions from such corporations are deemed to have caused subsequent damage via climate change to environmental or human health. For example, in 2008 the village of Kivalina in Alaska sued the Exxon Mobil Corporation, eight other oil companies and fourteen power companies for emitting greenhouse gases that, they claimed, contributed to the melting of sea ice, which

---

[27] See   www.cities-localgovernments.org/uclg/upload/news/newsdocs/climate_ agreement_ en.pdf [accessed 30 May 2008].

is threatening their community.[28] The relocation costs for the village were estimated at $400 million. Such unilateral actions invoke an instrument of the state – the legal system – as an instrument of climate governance.

Citizens. The third example of governance beyond the state is the development of Carbon Reduction Action Groups (or CRAGs) which are rooted in ideas of ecological citizenship. CRAGs are local community groups aimed at reducing the carbon footprints of members towards what they claim to be a sustainable and equitable level. They adopt the idea of voluntary personal carbon allowances, share knowledge and skills in lower-carbon living, and raise awareness and promote practical action in the wider community. The first CRAGs were established in the UK in 2006 and the movement has spread more recently to North America. CRAGs adopt the position that individuals need not accept the existing political and governance arrangements and can subvert these traditional arrangements through local action. By seeking to 'transform the relations between political power, consumption and global responsibility',[29] they are an example of what Portuguese cultural geographer Anabela Carvalho calls 'transformative global citizenship'.

Civic environmentalism therefore seeks co-operation between civil society, market and state to define and deliver climate governance. Multiple and diverse partnerships within and between nations and sectors offer greater scope for transformation than does over-reliance on UN Framework Convention negotiations between states and/or the

---

[28] CNN (2008) 'Climate change threatens existence, Eskimo lawsuit says', CNN News, 27 February. See www.edition.cnn.com/2008/WORLD/americas/02/26/us.warming.ap/index.html [accessed 30 May 2008].

[29] Carvalho, A. (2007) Communicating global responsibility? Discourses on climate change and citizenship. *International Journal of Media and Cultural Studies* 3(2), 180–3.

power of the market. Yet there are challenges to this eclectic approach to climate governance. One of these we have already encountered in Chapter 8: *The Challenge of Development* and is a radical form of the same ethos of civic environmentalism. Eco-anarchists and socialists such as the climate campaign Rising Tide challenge the institutions of the capitalist state, seeking a fundamental transformation of consumption patterns and existing institutions to pave the way for a more equitable and eco-centric world order. Such radicalism sits in a long tradition of political movements, such as the Second International at the end of the nineteenth century or the anarcho-syndicalists of the 1920s.

The other challenge is of a rather different type and has been most fully articulated by Australians David Shearman and Joseph Wayne-Smith in their book *The Climate Change Challenge and the Failure of Democracy*.[30] The problems of climate governance do not arise from too little democratic participation in public life – which the civic environmentalists might argue – but in fact they arise from too much democracy. The combination of a libertarian individualism, the hegemony of the market and the unchallenged pre-eminence of liberal democratic ideology – characteristics which now describe a large proportion of the Western world – makes it impossible for any such nation to drive forward the political, economic and social reforms that climate change mitigation might demand. This gloomy view of democracy sits comfortably with that expressed by British historian Eric Hobsbawm: 'Democracy, however, desirable, is not an effective device for solving global or transnational problems.'[31] For Shearman and Wayne-Smith the solution is the curtailment of individual liberties and the strengthening of the authoritarian centre. Civil freedoms and ecological survival, they argue, are in direct competition. This leads them to the thought that, 'it may well be non-Western states,

---

[30] Shearman, D. and Smith, J.W. (2007) *The climate change challenge and the failure of democracy*. Greenwood Press: Westport, CT.

[31] p. 118 in Hobsbawm, E. (2007) *Globalisation, democracy and terrorism*. Little, Brown: London.

including China, [who] will find ways to deliver [on climate change], while the West continues to display its extreme liberty with ineffectual debate and a surrender to powerful interests in its grinding democratic institutions'.[32]

These twin challenges to civic environmentalism – from the eco-anarchists and from the eco-authoritarians – may offer dismaying alternatives for many. They are representative of a new eco-political pessimism which climate change and our attempts to govern global climate have fostered. Whether these critiques are at all reconcilable with each other is far from clear, and whether China is a good example to be offered for strong climate governance is also questionable (see Box 8.2). These twin challenges offer visions for effective climate governance which are far removed from the conventional political wisdom that favours governance through the Kyoto Protocol and/or through global carbon markets.

9.5 The Clumsiness of Climate Governance

The above discussion highlights an irony of global climate change and our attempts to govern it. At a time of deepening political and social fragmentation, we have presented ourselves with climate change as a global phenomenon seemingly demanding co-ordinated, if not collective, planetary-scale governance and management. The political ideologies of the Cold War, and their centripetal effects on our modes of governance, dissolved in the late 1980s. And with this undermining of one powerful legitimator of the post-War nation-state has come an overall weakening of the very idea of the sovereign territorial state. Eric Hobsbawm, astute commentator on the events of the twentieth century, sums it up: 'Two and a half centuries of unbroken growth in the power, scope, ambitions and capacity to mobilise the inhabitants

[32] Shearman, D. (2007) 'Democracy and climate change: a story of failure', OpenDemocracy, 7 November. See www.opendemocracy.net [accessed 10 July 2008].

of modern territorial states, whatever the nature and ideology of their regimes, appears to be at an end.'[33]

At the same time, the increasing physical and virtual mobility of people and ideas facilitated by the internet and digital communications technologies further flattens and fragments the global political landscape. It is not surprising, therefore, that the unsettling challenges of climate change call forth extreme voices: a return to the authoritarian certainties of the ideological nation-state of the twentieth century, on the one hand, and the rise of a new radical civic environmentalism to finally overthrow the declining and antiquated powers of the compromised capitalist state, on the other.

We seem ill-equipped to bring an unruly climate and its implicated global citizenry under an appropriate form of governance. The nation-state is found wanting in the new complexities of multilateral climate diplomacy, the carbon market offers to share the burden of climate governance but engenders deep suspicion as to its motives, and a new civic environmentalism is caught between the horns of radicalism and conformism. Rather than a crisis of the environment or a failure of the market, climate change may prove ultimately to be a crisis of governance.

Before we leave this chapter we must visit two further dilemmas of climate governance, two challenging questions around which disagreement festers and is likely to grow: 'Will the Kyoto Protocol and its later derivatives ever be capable of delivering effective climate governance?' and 'Who governs the new geo-engineers?'

We have already sketched out the circumstances in which the Kyoto Protocol came into being in 1997, was ratified in 2005 and is being renegotiated for 2012. And we have already alluded to some of its critics and the challenges ahead if it is to harness the legitimacy of the nation-state, the power of the market and the vibrancy of civil society in a unifying global treaty. Many are committed to making

---

[33] Hobsbawm, *Globalisation, democracy and terrorism*, p. 105.

sure the Protocol survives – in modified form at least – beyond 2012. But others are questioning whether now is the time to disinvest political capital from the Kyoto process and rethink how best the goals of climate governance can be met. These calls for redirection are not new. In 2001, shortly after the USA withdrew from the Protocol, American political analyst David Victor wrote about 'the collapse of the Kyoto Protocol', arguing that the failure was due in large part to the mechanisms chosen. 'International law is a poor mechanism for allocating emissions permits.'[34]

A more recent and thorough critique of the Kyoto Protocol has come from two British policy analysts, Gwyn Prins and Steve Rayner. Their claim is that Kyoto has failed in several ways, not just because global carbon emissions have accelerated since 1997 rather than slowed, but also because it has stifled debate about alternative policy approaches to securing the goals of climate governance. Their complaint is that 'as Kyoto became a litmus test of political correctness, those who were concerned about climate change, but sceptical of the top-down approach adopted by the Protocol were sternly admonished that "Kyoto is the only game in town" '.[35] Agreeing with Victor that Kyoto was based on the wrong type of instrument – 'the wrong agents exercising the wrong sort of power' – Prins and Rayner offer a series of alternative approaches for securing climate policy goals to the ones offered up by the Kyoto process: to use their metaphor, 'silver buckshot' rather than a 'silver bullet' (see Box 9.2).

The Kyoto Protocol, as an approach to governing global climate, emerges almost exclusively from a hierarchical view of life. It is consistent with thinking which in other wider contexts has spawned ideas of a World Environmental Organisation or even, in the eyes of American international lawyer Richard Falk, of 'one-world government'. For

[34] p. 109 in Victor, D.G. (2001) *The collapse of the Kyoto Protocol*. Princeton University Press, Princeton, NJ, and Oxford.
[35] p. 973 in Prins, G. and Rayner, S. (2007) Time to ditch Kyoto. *Nature* 449, 973–6.

those who have less faith in the hierarchical structures, rules and managerial competences of society – those who hold a different view of Nature and our 'ways of living' – the Kyoto Protocol triggers vituperative reactions. Hendrik Tennekes, the then research director at the Royal Dutch Meteorological Institute, used powerful rhetoric to reveal such scepticism. 'I am terrified by the hubris, the conceit, the arrogance implied by words like "managing the planet" and "stabilising the climate". Who are we to claim that we can manage the planet? We cannot even manage ourselves? In an ecosystem no one is boss, virtually by definition. Why are we, with our magnificent brains, so easily seduced by technocratic totalitarianism?'[36]

The 'silver buckshot' approach of Prins and Rayner undercuts the hubris that non-hierarchists such as Tennekes may detect in the language and goals of the UN Framework Convention and the Kyoto Protocol. The idea of approaching the challenges of climate governance through a series of diverse, multi-level and almost deliberately overlapping, and even partly contradictory, institutions and policies is echoed by others who appreciate the insights of Cultural Theory. It allows the voices of egalitarians and individualists to be heard, even the voices of fatalists, whose policy horizons may be very limited. By appealing to a wider variety of instincts and constituencies, such a bottom-up approach to climate governance may offer a greater prospect of delivery. 'Silver buckshot' is an example of a 'clumsy solution' to climate change governance – a mixture of policy styles, normative principles and ways of viewing life and the world – an idea that has been promoted by a number of political scientists and cultural geographers and sociologists in recent years.[37]

---

[36] Tennekes, H. (1990) A sideways look at climate research. *Weather* 45, 67–8.

[37] This idea of clumsiness is well outlined in the book edited by Verweij, M. and Thompson, M. (2006) *Clumsy solutions for a complex world: governance, politics and plural perceptions.* Palgrave Macmillan, Basingstoke, UK, and was earlier outlined, in relation to climate change, in Thompson, M., Rayner, S. and Ney, S. (1998) Risk and governance. Part II: Policy in a complex and plurally perceived world. *Government and Opposition* 33(2), 330–54.

Box 9.2: The Wrong Trousers

Policy analysts Gwyn Prins and Steve Rayner issued a critique and offered an alternative to the Kyoto Protocol in a 2007 report titled *The Wrong Trousers: Radically rethinking climate policy*.[38] They used the metaphor of a pair of automated 'techno trousers' in which the hapless cartoon hero Wallace becomes trapped in the eponymous Oscar-winning animated film. Their argument was that the Kyoto Protocol, rather than making life easier, has marched the world involuntarily to unintended and unwelcome places in its well-meaning efforts to govern climate. Climate change is 'not amenable to an elegant solution because it is not a discrete problem'. In attempting an 'elegant' universal and top-down solution for governing global climate, the Kyoto Protocol has misread the problem. Instead, Prins and Rayner argue that any approach to the governing of climate should comprise a variety of approaches, each operating in different realms, at different levels and with different types of policy instruments – this is 'clumsy governance'.

These authors offer at least six elements to this clumsy approach to governance. The first is to abandon the ideal of Kyoto's universalism and focus just on the big carbon emitters. Dialogues undertaken by the G8+5,[39] for example, or agreements such as the Asia–Pacific Framework for promoting low-carbon technology between the USA, China, Japan, South Korea, India and Australia, are less likely to get side-tracked or diluted than the excessive multilateralism of Kyoto.

---

[38] Prins, G. and Rayner, S. (2007) *The wrong trousers: radically re-thinking climate policy*. Joint Discussion Paper of the James Martin Institute for Science and Civilisation, Oxford, and the MacKinder Centre for the Study of Long-Wave Events, London School of Economics, October.

[39] The G8 nations are the USA, the UK, France, Germany, Japan, Italy, Canada and Russia; the G5 are a group of large developing economies: China, India, Mexico, Brazil and South Africa. The G8+5 group was founded in 2005 following the G8 Summit hosted by British Prime Minister Tony Blair.

Second, they argue that the power of the market for regulating carbon emissions at least cost cannot be enabled from the top down. They advocate the cultivation of a number of bottom-up, semi-spontaneous carbon markets, following the Madisonian approach to carbon trading argued by David Victor.[40]

Third, and deliberately complementing the efficacy of the market, is their call for a serious expansion of public research and development (R&D) into new and renewable energy technologies. In this, they echo the call of many climate change commentators such as Ban Ki-Moon, Al Gore, and the president of Britain's Royal Society, Lord Rees.

Fourth, they advocate a new focus for multi-scale approaches to adapting to the climate risks to which many people and communities are exposed. Adaptation to climate risks is not an alternative to reducing carbon emissions; it is an essential – and for many regions a prior – investment which need not be tied to participation in any particular global climate treaty seeking to reduce carbon emissions. Most adaptation investment is 'no-regrets' in the sense that welfare benefits accrue to the citizens affected irrespective of future climate change or future climate mitigation.

Fifth is their call to move towards a 'global federalism of climate policy', which they believe will emerge from the rubble of the Kyoto Protocol. Climate governance should be multi-scale and should not solely, or even primarily, be conducted through the nation-state.

The final, and overarching, argument of Prins and Rayner is that inelegant, but attainable, bottom-up approaches to climate governance are likely to be more effective than elegant, but impractical, top-down ones. This is a view shared by others who doubt the wisdom or efficacy of the universalism of the Kyoto Protocol. The one 'gigantic package deal' for a post-2012 extension of the Kyoto

---

[40] Victor, D.G., House, J.C. and Joy, S. (2005) A Madisonian approach to climate policy. *Science* 309, 1820–1.

Protocol, which is demanded of the international negotiations to take place in Copenhagen in December 2009, is beyond our reach. A series of smaller, more flexible, multi-level and partly overlapping deals among subsets of actors, on the other hand, is attainable and hence more desirable.

Another challenge for global climate governance is emerging from a different quarter. A growing number of scientists, policy commentators and even private investors are considering the attractions of bypassing the demanding goal of mitigating climate change through a reduction in greenhouse gas emissions. Instead, they are developing technologies to enable the deliberate engineering of the world's climate. We introduced these so-called 'geo-engineering' technologies in Chapter 1: *The Social Meanings of Climate* and demonstrated their ideological appeal. What we need to consider here is whether and how schemes for geo-engineering global climate could effectively be governed.

Some argue that geo-engineering technologies can cut through the diplomatic bottlenecks which slow down and dilute the effectiveness of international treaties such as the Kyoto Protocol. A technology such as the direct capture of carbon dioxide from the free atmosphere[41] and its sequestration in underground reservoirs can be implemented anywhere in the world without the need for international negotiation. Enough carbon-sucking machines installed in one sovereign jurisdiction, for example, could in effect operate as a temperature regulator for the planet as a whole. These machines might be funded either publicly or, with a suitable price for tradable carbon permits,

[41] This technology has been developed by Columbia University physicist Klaus Lackner and a demonstrator prototype has been constructed by technology R&D company Global Research Technologies. The technique uses chemical processes to extract carbon dioxide from free-flowing air, and then buries underground the resulting solidified carbon compound.

through private investors. Other geo-engineering technologies – such as injecting aerosols into the stratosphere, installing mirrors in low Earth orbit at the Lagrangian point, or fertilising the oceans with iron[42] – offer similar unilateral opportunities for climate regulation.

Yet the implications of such actions undertaken outside any international agreement are legion. American atmospheric scientist Alan Robock has pointed out some of the potential sources of disagreement. 'How would the world agree on the optimal climate? What if Russia wanted it a couple of degrees warmer, and India a couple of degrees cooler? Should global climate be reset to pre-industrial temperature or kept constant at today's reading? If we proceed with geo-engineering, will we provoke future climate wars?'[43] In such a latter eventuality, climate change may yet be the subject of UN Security Council intervention, but in a different context to that envisaged in the debate with which we opened this chapter!

Given such considerations, one may argue that governing global climate through conventional approaches to emissions reductions may represent the easier challenge. And this is before one considers other dimensions of the Pandora's Box that geo-engineering opens up. There is already a history of states seeking to use deliberate weather modification for military purposes, notably the USA seeking to induce heavy rain to disrupt Viet Cong supply lines during the Vietnam War. For this reason and others, the UN Convention on the Prohibition of Military or Any Other Hostile Use of Environmental Modification Techniques has been ratified by eighty-five nations.

---

[42] Private organisations have been making plans to conduct large-scale iron releases in the Pacific Ocean to generate carbon offsets. This is despite significant uncertainties about the marine ecological effects of such releases; many scientists believe there is, as yet, no scientific basis for issuing carbon credits for ocean iron fertilisation. Buesseler, K.O. et al. (2008) Ocean iron fertilization – moving forward in a sea of uncertainty. *Science* 319, 161.

[43] p. 17 in Robock, A. (2008) Twenty reasons why geo-engineering may be a bad idea. *Bulletin of the Atomic Scientists* 64(2), May/June, 14–18, 59.

Could engineering the world's climate be effectively governed to prevent such hostile applications?

Geo-engineering also raises novel ethical dilemmas for climate governance. Many ethicists would argue that deliberately seeking to alter the world's thermostat, with consequences that are not fully predictable, is a categorically different ethical judgement than seeking to reverse or contain the consequences of what, historically, has been an inadvertent human modification of global climate. Geo-engineers are likely to be found among those who have the highest faith in models of the Earth system and among those who have the greatest confidence in the institutions of global governance. The former group believe that any unforeseen side-effects of geo-engineering are minimal and controllable, while the latter believe that such technologies end up serving the global good, rather than sectarian interests. Advocates of such purposeful large-scale interventions with climate are more likely to be found among hierarchists and individualists than among egalitarians.

9.6 Summary

So why do we disagree about climate change? We have shown in this chapter that the dominant attempts to establish climate governance – to design the architecture of global climate policy – have relied heavily on the political authority of the nation-state and on the economic efficiency of the market. The Kyoto Protocol and various forms of international carbon trading form the framework around which most political assent and dissent has congregated.

Yet few commentators or analysts seem satisfied with these tentative steps towards global climate governance. Frustration and failure seem to have characterised the years since 1997. If the success of these tentative steps is measured in terms of reductions in global greenhouse gas emissions or in the additional numbers of people protected against climate risks then this dissatisfaction seems warranted.

On the other hand, the defenders of these moves towards climate governance – towards an exercise of authority over climate – might point to the very existence of the Kyoto Protocol as its own self-referencing criterion of success. They might also point to the creation of an internationally recognisable price for carbon.

But we have shown that climate governance is not simply about the Kyoto Protocol, the nation-state and the carbon market. Moves to 'govern' climate are in evidence within and across societies, through initiatives, movements and alliances which seek to bypass the formalities of political negotiations and carbon trading. 'Action on climate change' is the mantra of this new civic environmentalism, actions which take many forms: citizen's pledges, technology standards, voluntary carbon offsets, legal prosecutions. This eclectic and diverse form of civic action – and the tension this implies between top-down and bottom-up governance – helps to explain why public surveys repeatedly reveal confusion among citizens over who is perceived to be responsible for 'taking action' on climate change: the United Nations, governments, businesses, the rich, or individual citizens.

This chapter has also revealed deeper reasons why we disagree about how climate may be governed. These reasons emerge not only from our different ways of life – the hierarchists' preference for top-down universal regimes or the egalitarians' preference for more inclusive and participatory approaches. They also emerge from the exercise of political power, from the actions of nation-states seeking (economic) self-interest. These actions have a history. They emerge from different historical and ideological traditions of the relationship between state and citizen and from different interpretations of the historical relationships between North and South. Liberal democracies approach the governance of climate differently from authoritarian regimes. And the powers of Europe, with their early industrialisation and imperial histories, bring a different historical legacy into climate negotiations than do their former colonies, now emancipated and seeking redress for past injustices. And the emerging global superpowers of China and India see concerns about climate change as an opportunity to demonstrate that a new world political order is emerging in a new century.

Who then governs global climate? And in whose interests do they govern? It is naive to think that any of the institutions of science, economics and religion we examined earlier in the book can offer an ultimate blueprint for the governing of global climate. And we have seen, too, how the different perceptions and representations of climate risks, and their convoluted relationships with the multiple ideas of sustainable development, further complicate the political landscape. In seeking to govern climate, humanity is attempting something that has never before been attempted – global governance of the global atmosphere. This is akin to the first experiments in democracy in Athens in the fifth century BC, the establishment of British parliamentary democracy in the seventeenth century, or the creation of the League of Nations/United Nations in the twentieth century.

Stating the challenge of this new governance project, and the idealised requirements to meet it, is easy. Timmons Roberts and Bradley Parks, for example, in their book, *A Climate of Injustice*, make it sound almost plausible. 'Aiding nations in making the difficult transitions to more equitable and economically sustainable and lower-carbon pathways of development may be the only way to resolve the issue of climate change. But this transition needs to be built on generalised reciprocity, a climate of trust, shared principles of justice and a common world-view of environment and development.'[44] But building trust between nations, agreeing shared principles of justice, and agreeing a common view of environment and development cannot be delivered through wishful thinking. We have seen in this chapter, and in the earlier ones, how difficult such agreement is. Politics may be the ultimate social process we have yet designed for groups of

[44] p. 24 in Roberts, J. T. and Parks, B. C. (2007) *A climate of injustice: global inequality, North–South politics and climate policy*. MIT Press: Cambridge, MA.

people to reach, legitimate and enforce collective decisions, but the global politics of climate change tests that process more than it has been tested before.

So what hope is there for us? Does our story about climate change have to end in the quagmire of global politics? We will see in Chapter 10: *Beyond Climate Change* that there are other ways of thinking about climate change which do not require unreachable global agreements; ways of thinking and telling stories about climate change that take us in different directions. Rather than being approached as a problem to be solved, climate change can become an imaginative resource to be used.

FURTHER READING FOR CHAPTER 9

Aldy, J. E. and Stavins, R. N. (eds) (2007) **Architectures for agreement: addressing global climate change in the post-Kyoto world.** Cambridge University Press.
This edited book examines six alternative international architectures for establishing global climate policy. An advocate first presents each proposal, with two more critical contributors then responding. This is one of the more accessible and wide-ranging books on different approaches for governing climate through the nation-state and international agreements, with scholars contributing from economics, law and international relations.

Benedick, R. (1998) **Ozone diplomacy: new directions in safeguarding the planet** (2nd edn). Harvard University Press: Cambridge, MA.
This book is frequently cited as the definitive book on the most successful global environment treaty and how it was negotiated and implemented. Benedick offers an upbeat insider's account of how this global governance regime for stratospheric ozone emerged. In his final chapter, he offers an analysis applying the lessons of the ozone experience to climate change negotiations.

Newell, P. (2006) **Climate for change: non-state actors and the global politics of the greenhouse.** Cambridge University Press.
In contrast to Aldy and Stavins, this book shows how it is not just the interests of the nation-state that influence global climate politics. Newell assesses and explains the influence that non-state actors – scientific, environmental, civic and business groups and movements – have brought to bear on the course of global politics. The book offers a good introduction to the complexity of the politics of climate change, and the limits of the power of national governments.

Pettenger, M. (ed.) (2007) **The social construction of climate change.** Ashgate Press: Aldershot, UK.
This collection of essays examines different facets of the policy and governance dilemmas posed by climate change and seeks to explain them by understanding the contextual relationships between power, knowledge and society. Together, they help to reveal why climate change has produced so many seemingly intractable disputes about the goals, processes and implementation of climate policy.

Shearman, D. and Smith, J. W. (2007) **The climate change challenge and the failure of democracy.** Greenwood Press: Westport, CT.
In this provocative book, Shearman and Smith present evidence that the fundamental problem causing environmental destruction – and climate change in particular – is the operation of liberal democracy. Its flaws and contradictions bestow upon government – and its institutions, laws, and the markets and corporations that provide its sustenance – an inability to make decisions that could provide a sustainable society.

TEN

Beyond Climate Change

10.1 Climate Change is Everywhere

Climate change is everywhere. Not only are the physical climates of the world everywhere changing, but just as importantly the *idea* of climate change is now to be found active across the full parade of human endeavours, institutions, practices and stories. The idea that humans are altering the physical climate of the planet through their collective actions, an idea captured in the simple linguistic compound 'climate change', is an idea as ubiquitous and as powerful in today's social discourses as are the ideas of democracy, terrorism or nationalism. Furthermore, climate change is an idea that carries as many different meanings and interpretations in contemporary political and cultural life as do these other mobilising and volatile ideas.

Climate change, then, is to be found everywhere. Just as the transformation of the world's physical climates is inescapable, so is the idea of climate change unavoidable. It is an idea circulating anxiously in the worlds of domestic politics and of international diplomacy. It is an idea circulating with mobilising force in the worlds of business, of law and of international trade. It is an idea circulating with potency in the worlds of knowledge and invention, of development and welfare, of religion and ethics, and of public celebrity. And it is an idea circulating creatively in the worlds of art, of cinema, of literature, of music and of sport.

Evidence for this penetration of climate change into these social and imaginative worlds is not hard to find. As I was writing this chapter in Norwich, England, during one week in May 2008, the following announcements, reports, events and stories passed across my computer screen – multiple worlds, networks and movements in my own country connected uniquely through their engagement with and deployment of the idea of climate change. Consider the following items:

- A new interdisciplinary network was announced at the University of Cambridge, exploring the issues surrounding **climate change** from an aesthetic and cultural standpoint, giving 'voice and platform to the growing number of artists, writers, photographers, journalists, and film directors responding to this global phenomenon'.
- The Royal International Institute of International Affairs in London held a one-day seminar on '**Climate change** and intellectual property rights'.
- The UK Government's initiative 'Together We Can', a campaign to bring government and people closer together, released an eponymous film featuring well-known figures from government, business and entertainment – Tony Blair, Stuart Rose, Annie Lennox, Claudia Schiffer – uniting behind one simple message: 'By working together we can make a real difference in the fight against **climate change**.'
- The European Commission and the University Association of Contemporary European Studies organised a two-day conference on '**Climate change**, energy security and Europe's next big project'.
- Birmingham City Council announced the world's first festival to link **climate change** with urban planning and design, with the goal of 'transforming the quality of life for people working and living in the city'.

- The New Economics Foundation in London published a report *The Response of Civil Society to* **Climate Change**, exploring the relationship between civil society, social justice and climate change.
- The UK Women's Research and Enterprise Forum held a workshop at the University of East Anglia exploring new business opportunities for female entrepreneurs offered by **climate change** and the low-carbon economy.

And the insinuation of climate change into our social worlds extends well beyond the UK. During this same week in May 2008 the idea of climate change also announced itself to me through the globalised webs that connect North with South, East with West and through which flow financial, material, virtual and intellectual goods. Consider the following items:

- A new five-year exhibition introducing environmental challenges facing the world was announced by the Science Centre Singapore, together with the National Environment Agency and Shell Companies in Singapore. **Climate change** was unveiled as one of its major exhibitions for 2008.
- The Second International Conference on '**Climate change** as a challenge for poverty reduction' took place in Bonn, Germany, claiming that 'climate change poses a significant threat to the goals of the global fight against poverty'.
- A study commissioned by the Andean Community of Nations and carried out by the Peruvian University of the Pacific reported that **climate change** could cost Andean countries US$30 billion per year by 2025. Peru's former agriculture minister and co-ordinator of the research team stated that this is equivalent to 'the amount that [the Andean] countries devote to public health expenses today'.
- A Sudanese climate researcher, Balgis Osman-Elasha, was honoured by the UN Environment Programme in recognition of her work on **climate change** and adaptation in conflict-stricken Darfur. She was presented with a 'Champions of the Earth 2008' award for

her work on how communities in Darfur could cope with drought, saying: 'We should act now and curb climate change.'

- The *Washington Post* announced that some of the world's major agri-biotech companies were applying for hundreds of patents on genetically engineered 'climate crops', carrying out what amounts to an 'intellectual property grab' in this new lucrative **climate change** market.

It is not sufficient to argue that these events, initiatives, reports and innovations are connected to each other simply through each being an instinctive and uncomplicated response to the transformation of the world's physical climates. The adoption of the phrase 'climate change' mobilises in each case very different sets of ideologies, meanings, values and goals. And the diversity of meanings attached to this expression expands further if one moves outside these formalised webs of exchange, into communities less exposed to globalised knowledge, politics and entertainment. These mobilising narratives cannot be found in the discoveries of science; they cannot be read from the scripts of the scientific assessments of the UN Intergovernmental Panel on Climate Change (IPCC). The idea of climate change means different things to different people in different contexts, places and networks.

There is no single perspective or vantage point from which this kaleidoscopic idea of climate change can be understood. The IPCC – which arguably first introduced the physical reality of climate change into these social worlds – has constructed and presented a powerful scientific consensus about the physical transformation of the world's climate. This is a reality I believe in. But engaging with climate change takes us well beyond the physical transformations that are observed, modelled and predicted by natural scientists and assessed by the IPCC. Science may be solving the mysteries of climate, but it is not helping us discover the meaning of climate change. We need new ways of looking at the phenomenon – an idea circulating and mutating through our social worlds – and of making sense of what climate change means to us. One way of expressing this distinction between climate change as physical transformation and climate change as an idea is offered in Box 10.1.

This final chapter of *Why We Disagree About Climate Change* offers a perspective on climate change which transcends the categories and disagreements explored in earlier chapters. It looks beyond the disputes about scientific knowledge and whether a particular weather event was caused by global warming. It looks beyond complaints about whether this particular TV programme or that particular newspaper article got the science 'right'. It looks beyond the debates about the most suitable discount rate to be used in economic analyses of climate change. And it looks beyond arguments about whether this particular country or that particular president has helped or hindered the 'fight' against climate change.

In this chapter I argue that climate change is not a problem that can be solved in the sense that, for example, technical and political resources were mobilised to solve the problem of stratospheric ozone depletion. Instead, I suggest a different starting point for coming to terms with the idea of climate change. I believe that human beings are more than material objects and that climate is more than a physical category. I suggest we need to reveal the creative psychological, ethical and spiritual work that climate change is doing for us. Understanding the ways in which climate change connects with foundational human instincts opens up possibilities for re-situating culture and the human spirit at the heart of our understanding of our changing climate. Rather than catalysing disagreements about how, when and where to tackle climate change, the idea of climate change should be seen as an intellectual resource around which our collective and personal identities and projects can form and take shape. We need to ask not what we can do for climate change, but to ask what climate change can do for us.

Box 10.1: Lower-Case and Upper-Case Climate Change

Some of the cultural and ideological meanings we have histori-cally attached to climate were explored in Chapter 1. I now want to suggest a way of distinguishing between two different senses in which the words 'climate change' may be used; a distinction that has been latent within much of what I have written. Borrowing an idea from intellectual history,[1] I suggest that a distinction between a lower-case and an upper-case reading of climate change might be helpful.

Reconstructionist lower-case history refers to the practice of his-tory in which historians believe they are retelling 'the past as it really was'; a reconstruction uncomplicated and untainted by ide-ology. Lower-case climate change might therefore refer by analogy to the transformation of physical climates; a phenomenon which scientists study using the 'objective' and impartial methods and tools of science. Conversely, constructionist upper-case History is the past retold and reinterpreted through the lens of one or more committed ideological positions, for example Marxist History, Imperial History or New Economic History. Upper-case Climate Change might thus refer by analogy to the phenomenon of a physi-cally transformed climate, but now entangled with and interpreted by the ideologies of the stories that are told about Climate Change and what it signifies.

To illustrate this nomenclature, consider the different ways in which 'climate change' is used in the following two publications. Meteorologists Bradley Murphy and Bertrand Timbal published a review paper in the *International Journal of Climatology* in June 2008 called 'A review of recent climate variability and climate

---

[1] For example, see Keith Jenkins, writing in the Introduction to Jenkins, K. (ed.) (1997) *The postmodern history reader*. Routledge: London.

change in southeastern Australia'. I suggest that we should think of this as an example of lower-case usage, 'climate change' as a noun, in which the changing physical properties of climate over time were investigated using scientific categories. On the other hand, geographer John Robinson and colleagues produced, also in 2008, a special issue of the journal *Climate Policy* called 'Integrating Climate Change Actions into Local Development'. I suggest that we should think of this as an example of upper-case usage, 'climate change' used here as an adjective to describe a certain category of individual and collective behaviours. What actions qualify as 'climate change actions', as opposed to other types of actions, depends on an array of assumptions and ideological filters which renders the idea of Climate Change in this case very different from the climate change object being studied by Murphy and Timbal. Indeed, the relationship between these two categories of meaning is not at all self-evident.

This distinction between climate change as noun and adjective, between climate change as physical phenomenon and as an idea, may not always be unambiguous. But as stylised caricatures of two extreme positions it helps to expose the cause of some of our confusions about climate change. Later in this chapter, for example, I explicitly discuss Climate Change rather than climate change.

There are serious limits to the 'problem–solution' framing of climate change and these are explored in Section 10.2. What was being constructed by climate scientists in the 1980s and 1990s as an environmental problem has turned into something very different. I argue that climate change has more potency now as a mobilising idea than it does as a physical phenomenon. Ideas can be used, but they can't be solved. Climate change can no longer be approached as an environmental problem demanding technical solutions. Climate change is not like lead in petrol or asbestos in construction – undesirable physical substances to be eliminated or regulated. Neither is climate change simply a social problem seeking a political solution. It is not like slavery or domestic violence; distortions of human relationships to be outlawed and policed. Climate change will not be 'solved' by science or technology – although science will continue to yield new insights for us about our changing climates, and technology will continue to offer us new ways of living within them. But neither will climate change be 'solved' by politics or economics – although politics will seek pathways to co-operative action and economics may soften some of the hard edges of our new life with climate change. Section 10.2 uses the concepts of 'wicked problems' and 'clumsy solutions' to develop this argument, and shows why even these more subtle ways of approaching climate change through a 'problem–solution' framework are unlikely to deliver.

Rather than try to 'solve' climate change, I suggest in Section 10.3 that we need to approach climate change as an imaginative idea, an idea that we develop and employ to fulfil a variety of tasks for us. Because the idea of climate change is so plastic, it can be deployed across many of our human projects and can serve many of our psychological, ethical and spiritual needs. This section offers four mobilising narratives of climate change; narratives rooted in our human instincts for nostalgia, fear, pride and justice. I attach metaphors to these stories – respectively labelled lamenting Eden, presaging Apocalypse, constructing Babel, and celebrating Jubilee – about how the idea of climate change serves our human needs.

Since I argue that the ultimate significance of climate change is ideological and symbolic rather than physical and substantive, Section 10.4 returns us to Chapter 1: *The Social Meanings of Climate*. There I argued that our sensual experiences and scientific depictions of physical climates have always been inextricably – if often unreflexively – entangled with meanings that reflect cultural and ideological movements. Here I argue that in the twenty-first century it is climate change and not climate *per se* that is becoming the new locus of such entanglements. Just as it is important to recognise that the idea of *climate* cannot be fully 'purified' of our cultural and personal imaginations, so now we must realise that the idea of *climate change* demands interpretation through these same imaginative faculties. Science has universalised and materialised climate change; we must now particularise and spiritualise it. The four stories I offer of Eden, Apocalypse, Babel and Jubilee are attempts to do just this by holding up mirrors in which we can see ourselves reflected and which reveal some of the contradictions that make us human.

The chapter, and the book, concludes by looking beyond climate change. The world's climates will keep on changing, for our generation and for generations to come. Human influences on the physical properties of climate are now, and probably will be forever, inextricably entangled with natural forces. But so too will the *idea* of climate change keep changing as we find new ways of using it to meet our needs. We will continue to create and tell new stories about climate change and mobilise them in support of our projects. These stories may teach us to embark on different projects. And we will disagree. But we can at least recognise that the sources of our enduring disagreements about climate change lie within us, in our values and in our sense of identity and purpose.

10.2 Why Climate Change Will Not Be Solved

The Fourth Assessment Report of the IPCC in 2007 was one in a series of events, reports, statements and campaigns over recent years which have each been heralded as marking *the* turning point in the fight against climate change. Each was a moment when commentators offered us the prospect that solutions to climate change would, at last, become more imminent and salient than the causes of climate change. We can mention a few of them.

On the publication of the Stern Review in October 2006, Oxford economist Cameron Hepburn claimed that 'when the history of the world's response to climate change is written, the Stern Review will be recognised as a turning point', while Claude Mandil, Executive Director of the International Energy Agency, congratulated 'Sir Nick Stern and his team for producing a landmark review which I have no doubt will strengthen the political will to change of governments around the world'.[2] A few months later, in February 2007, when the IPCC published its Fourth Assessment Report on the science of climate change, similar optimism about the transition from argument to action was being voiced. 'UN's vast report will end the scientific argument. Now will the world act?'[3] This was a sentiment shared a few weeks later by then UK Environment Secretary David Miliband, as he returned from a meeting of the Global Legislators Organisation for a Balanced Environment (GLOBE) in Washington: 'This year could potentially be a tipping point in the fight against climate change.'[4]

The award of the Nobel Peace Prize jointly to Al Gore and the IPCC in October 2007 saw a further surge of optimism that a turning point had been reached; Governor Bill Richardson of New Mexico claiming, 'Al Gore's work on global warming has been a turning point in this battle, and his award today is a fitting recognition'.[5] And the agreement of the Bali Action Plan in December 2007 at COP-13, setting out the steps for achieving a post-2012 follow-on to the Kyoto Protocol, was also greeted effusively. UN Secretary-General Ban Ki-moon opened the Bali Conference with the words: 'We gather because the time for equivocation is over. Climate change is the defining challenge of our age. The science is clear; climate change is happening, the impact is real. The time to act is now.' The Indonesian Environment Minister and President of the conference, Rachmat

[2] Comments on the Stern Review, HM Treasury. See www.hm-treasury.gov.uk/ independent_reviews_index.htm [accessed 12 June 2008].
[3] *Guardian*, 27 January 2007.
[4] David Miliband, quoted in *New Scientist*, 24 February 2007.
[5] Quoted 12 October 2007, www.richardsonforpresident.com [accessed 14 November 2008].

Witoelar, applauded the outcome: 'This is a real breakthrough, a real opportunity for the international community to successfully fight climate change.'[6]

These voices spoke from the worlds of politics, science, economics and journalism. They have been the voices of optimism and hope around which policy advocates, campaigners, business and religious leaders, citizens and celebrities have congregated and further amplified and transmitted the message to the far reaches of our societies. 'The time to act is now.' These siren voices have explained, cajoled, scolded, bullied, preached – pleaded even – society to take heed, to repent, to change beliefs, behaviours and practices.

But enlightenment has not arrived. The world has remained stubborn and has not turned. Some policy innovations have occurred here and there. Some businesses have taken the plunge into carbon markets. Direct action groups have mobilised against airport expansion and coal-fired power stations. And many millions of the masses have signed up for voluntary emissions reduction pledges. But emissions of greenhouse gases keep on rising – globally by 16 per cent in the decade since the Kyoto Protocol was first negotiated – and so does the world's temperature and sea level.

We need to face up to an uncomfortable reality. We have lived through more than twenty years of IPCC assessments, more than sixteen years of the UNFCCC negotiations, more than a decade of activities inspired by the Kyoto Protocol (several years of which have seen the Protocol in full force) and successive rounds of G8 conferences at which climate has been top of the agenda: but none of these global deployments of science, economics, international relations, diplomacy and politics have yielded the prize being sought. Senator Richardson's 'turning point' has *not* been reached; Minister

---

[6] Quoted in 'UN climate conference adopts plan to negotiate new global warming pact by 2009', *China View*, 15 December 2007. See http://news.xinhuanet.com/ english/2007-12/15/content_7254553.htm. [accessed 14 November 2008].

Witoelar's 'breakthrough' has *not* come. Earth system science has told us that the physical climate system is sluggish to external forcing. We have now also discovered that the world's energy economy is sluggish with respect to technological innovation and – no discovery this – that human behaviour is resistant to mere exhortation.

Perhaps this particular way of framing climate change (as a mega-problem awaiting, demanding, a mega-solution) has led us down the wrong road. By constructing climate change as the 'mother of all problems' – 'the [greatest/defining/most serious]* long-term [problem/challenge/threat]* facing humanity' – perhaps we have out-manoeuvred ourselves. We have allowed climate change to accrete to itself more and more individual problems in our world – unsustainable energy, endemic poverty, climatic hazards, food security, structural adjustment, hyper-consumption, tropical deforestation, biodiversity loss – and woven them together using the meta-narrative of climate change.

We have created a political log-jam of gigantic proportions, one that is not only insoluble, but one that is perhaps beyond our comprehension. Combine this with the rhetoric of fragility and disaster – 'ten years to avoid catastrophic tipping points'[7] – and our post-modern 'disinclination towards meta-narratives',[8] and perhaps we really are already doomed. No wonder the planetary geo-engineers are emerging from the shadows offering us their alternative Promethean visions of the 'mother of all solutions'.

* Delete as desired

[7] Extract from a letter to European leaders in October 2006 from UK and Dutch Prime Ministers Tony Blair and Jan Peter Balkenende. The relevant sentence was, 'We have a window of only ten to fifteen years to avoid crossing catastrophic tipping points.'

[8] Jean-Francois Lyotard's seminal 1979 essay – published in 1984 as *The postmodern condition: a report on knowledge*. University of Minnesota Press: Minneapolis, MN – suggested that this was a defining characteristic of post-modern societies, a scepticism towards traditional meta-sources of knowledge and meaning.

Climate change as a 'wicked problem'

More than thirty years ago, planning theorist Horst Rittel proposed the term 'wicked problems' to describe a category of public policy concerns that defied rational and optimal solutions. For Rittel and his colleague Melvin Webber,[9] wicked problems are essentially unique, have no definitive formulation, and can be considered symptoms of yet other problems. Solutions to wicked problems are difficult to recognise because of complex interdependencies in the system affected; a solution to one aspect of a wicked problem often reveals or creates other, even more complex, problems demanding further solutions.

This description applies very well to what is conventionally understood as the 'problem' of climate change. As our earlier chapters have shown, climate change, as conventionally framed, possesses all the attributes of a wicked problem, a situation defined by 'uncertainty; inconsistent and ill-defined needs, preferences and values; unclear understanding of the means, consequences or cumulative impacts of collective actions; and fluid participation in which multiple, partisan participants vary in the amount of resources they invest in resolving problems'.[10] 'Tame' problems, on the other hand – while they may be complicated – have relatively well-defined and achievable end-states and hence are potentially solvable. The example of stratospheric ozone depletion (see Box 9.1) may fall into this category. Wicked problems, however, afflict open, complex and imperfectly understood systems, and are beyond the reach of mere technical knowledge and traditional forms of governance.

Failing to understand and treat climate change as a 'wicked problem' has led to the construction of a global solution-structure that possesses elements that appear either inadequate or inappropriate

[9] Rittel, H. and Webber, M. (1973) Dilemmas in a general theory of planning. *Policy Sciences* 4, 155–69.
[10] p. 156 in Carley, M. and Christie, I. (2001) *Managing sustainable development* (2nd edn). Earthscan: London.

given the intractability of climate change. We have described and commented on a number of these elements in earlier chapters, pointing out how often they either act as attractors for vigorous disagreement or else simply transfer the problem somewhere else. We repeat a few of them here:

- The desire to establish a single universal policy target, namely a global atmospheric concentration of greenhouse gases target which putatively avoids 'dangerous' climate change (see Chapter 3: *The Performance of Science*).
- The desire to establish a single carbon market with worldwide trading, thereby correcting the 'market failure' which, it is claimed, underpins climate change (Chapter 4: *The Endowment of Value*).
- The desire to effect a social revolution from the 'bottom-up', thereby rethinking ideas of consumption, growth and capitalism (Chapter 5: *The Things We Believe* and Chapter 8: *The Challenges of Development*).
- The desire to minimise poverty worldwide, thereby reducing vulnerability, social injustice and political instability (Chapter 8).
- The desire to move research and development (R&D) investment in zero-carbon energy technology onto a 'wartime' footing, thereby securing technological solutions to climate change (Chapter 9: *The Way We Govern*).
- The desire to establish a single global policy regime or protocol to which all nation-states are signatories, thereby establishing a means of global climate governance (Chapter 9).
- The promotion of a new generation of geo-engineering technologies, thereby countering the additional greenhouse warming without relying on any of the above, namely: carbon markets, clean technologies, redistributive growth and consumption management (Chapter 9).

The problem with all these different, and sometimes contradictory, elements of the global solution-structure is that they underestimate the 'wickedness' of climate change and overestimate the abilities of economics, politics or technologies to tame and master our changing climate. These solution elements rely too heavily on either rational choice theory in economics, regime theory in politics, social coercion in behaviour management, or control engineering in the implementation of technology.

This global solution-structure also begs a fundamental question which is rarely addressed in the respective fora where these debates and disagreements surface: What is the ultimate performance metric for the human species, what is it that we are seeking to optimise? Is it to restabilise climate, to stabilise population or to minimise our ecological footprint? Is it to increase life expectancy, to maximise gross domestic product (GDP), to make poverty history or to increase the sum of global happiness? Or is the ultimate performance metric for humanity simply survival?

The ultimate goal of humanity may be clear to those who have elevated restabilising climate to the top of their list, but wicked problems do not yield to such one-dimensional goal-setting. And setting the overarching goal of humanity as the restabilisation of climate will, I believe, lead to disillusionment. Carbon dioxide concentrations will continue to rise, world temperature to warm, global sea level to rise, and weather to wreak social havoc. These headline indicators of climate change will resist our interventions for at least another generation, if not longer, irrespective of our successes in slowing the growth in global greenhouse gas emissions. We will see no quantitative indexed evidence of the 'solution' for at least half a lifetime or more.[11]

---

[11] Owing to the time lags between reducing emissions and reducing concentrations, between reducing concentrations and reducing temperature, and between reducing temperature and reducing sea-level, even if the world *was* to succeed in first slowing the growth of emissions and then reducing the absolute level of emissions, the visible effects of these achievements on these indicators of the climate system, relative to today's conditions, will take at least a generation to emerge, most probably much longer.

Or imagine a world fifty years from now, in the year 2060, in which optimistically the global greenhouse gas concentration has stabilised at between 450 and 550 ppm. There may well then be prospects for the global temperature to rise not much above 2° or 3°C beyond the pre-industrial value and for the eventual rise in global sea level to be contained to less than 1 metre. In this imagined future, we may have come close to securing our primary goal of restabilising global climate, but will the world be a better place? In 2060 we will still live in a world with wars, poverty, inequality, hunger and disease. Cyclones, droughts and floods will continue to cause death and destruction. The mere restabilising of climate need not necessarily help in reducing any of these material afflictions relative to today's experiences.[12] It is difficult to see any reason why, in this future climate-stabilised world, our population will be lower, our ecological footprint smaller, our poverty reduced, our GDP greater, our life expectancy longer, or our happiness enriched relative to any other possible future world. What is the human project ultimately about?

Climate change as 'clumsy solutions'

One possible response to recognising that climate change is a wicked problem rather than a tame one is to rethink the nature of the solutions we seek and the way in which we implement them.[13] One adjustment could be to embrace the idea of 'clumsy solutions', an idea we introduced in Chapter 9: *The Way We Govern*. The term 'clumsy institutions' was coined by American law professor Michael Shapiro

---

[12] Of course, relative to some other hypothesised future worlds in which climate is not stabilised by 2060, all these indicators may be 'better'; but these improvements do not necessarily follow from securing the objective of stabilising climate.

[13] After completing the draft of this book, it was brought to my attention that wickedness and clumsiness had also been connected with regard to climate change by Steve Rayner in his 2006 Jack Beale Memorial Lecture, Sydney, Australia: 'Wicked problems: clumsy solutions: diagnoses and prescriptions for environmental ills'. See: www.martininstitute.ox.ac.uk/JMI/Library/James+Martin+Institute+Editorial/Jack+Beale+Memorial+Lecture+2006.htm [accessed 25 November 2008].

in 1988 as a way of escaping from the idea that, when faced with contradictory definitions of problems and solutions, only *one* definition must be chosen and all others rejected. Clumsiness allows for several or all such contradictory goals and policies to be simultaneously pursued.

Clumsiness therefore emerges as the opposite of elegance or optimality in policy making. It sits uneasily alongside universalist mentalities, whether those inspired by science, economics, religion, risk management, development or politics. A belief in clumsy solutions demands that multiple values, multiple frameworks and multiple voices be harnessed together – clumsily, contradictorily – in our response to wicked problems. 'The Earth is one, but the world is not,' declared Gro Harlem Brundtland's *Our Common Future* in 1987. Earth system science may demand and find a unitary framework of explanation and prediction, but our social worlds resist such unifying frameworks.

The sagacious British historian, Isaiah Berlin, presaged this way of thinking many years ago in 1969 when he wrote in his essay *Liberty*: 'What the age calls for ... is not more faith, stronger leadership or more scientific organisation. Rather it is the opposite – less Messianic ardour, more enlightened scepticism, more toleration of idiosyncrasies, more frequent *ad hoc* measures to achieve aims in a foreseeable future.'[14] Berlin would have approved of the metaphor of 'silver buckshot' adopted by Prins and Rayner to describe possible clumsy solutions to climate change in preference to the metaphor of a 'silver bullet' (see Chapter 9). Clumsiness suggests that we construct our problems in such a way as to make them fit our capabilities for solution-making rather than imagine that our human ingenuity can find solutions to whatever problems we casually invent.

Clumsiness certainly has some attractions, but the limitation with thinking about climate change even in terms of clumsy, suboptimal

---

[14] p. 92 in Berlin, I. (2002) *Liberty* (revised and expanded edition of *Four Essays On Liberty*, 1969). Oxford University Press: New York.

solutions is that we are still framing climate change as a problem to be solved. *If* emissions start falling, *if* greenhouse gas concentrations stabilise, *if* global temperature and sea level are quantities brought under human regulation, then – and only then – will we have succeeded. We have made climate change the overriding project of this generation. We have been persuaded by our own rhetoric that it is on this project – and this project alone – that future generations will judge us. Veteran environmental campaigner George Monbiot has issued the rallying cry: 'Curtailing climate change must ... become the project we put before all others. If we fail in this task, we fail in everything else.' [15]

But clumsy solutions are not only suboptimal in design, they are suboptimal in outcome and their metrics are incommensurable. By attending to different facets of the more undesirable drivers and outcomes of climate change, and by doing so through a diversity of means and at different scales, clumsy solutions will not be contributing to the same overall performance measure. They may end up cancelling each other out. Success in expanding biofuel production to displace fossil carbon in our liquid fuel mix reduces biodiversity and food security – just the outcome from climate change we seek to avoid. Improving energy efficiency in consumer electronics rebounds to expand the use of energy through other consumption practices. Pricing internationally traded goods according to their carbon footprint undermines the livelihoods of subsistence farmers in Kenya and increases their exposure to the vicissitudes of climate. These perversities of outcome are characteristics of wicked problems that even clumsiness cannot eliminate.

The globally connected webs of relationships and dependencies that are traced through the flows of material, financial, social and virtual capital react and adjust in complex and unpredictable ways. As sociologist Bruno Latour has commented in the context of theories

---

[15] p. 15 in Monbiot, G. (2006) *Heat: how to stop the planet burning.* Allen and Lane: London.

of risk, the side-effects of our clumsy solutions to climate change may be 'so numerous that their proliferation has eaten up projects to the point where the projects themselves have been transformed beyond recognition'.[16] What start off as responses or solutions to climate change may end up redefining and perhaps multiplying the problem in novel ways. We must recognise the 'wickedness' of climate change and we must appreciate that while clumsiness – with all its contrariness and messiness – is perhaps the limit of our human ability to respond, it will not deliver the outcomes we seek.

But if this is all climate change means to us – a (wicked) problem continually awaiting (clumsy) solutions – we have limited our imaginations. There is a further position beyond climate change that we must reach.

10.3 Four Myths[17] of Climate Change

If climate change is not a problem that can be solved – either through elegant solutions or through clumsy ones – we need to find other ways of categorising it. In order to get a better vantage point I suggest we change our position and examine climate change as an idea of the imagination rather than as a problem to be solved. By approaching

---

[16] p. 38 in Latour, B. (2003) Is re-modernization occurring – and, if so, how to prove it? A commentary on Ulrich Beck. *Theory, Culture and Society* 20(2), 35–48.

[17] I use the term 'myth' in this chapter in the very specific anthropological and non-pejorative sense of revealing meanings and assumed truths. 'To talk about [myth] as truth is a little bit deceptive because when we think of truth we think of something that can be conceptualized. [A myth] goes past that' (p. 1 in Campbell, J. and Moyers, B. (1988) *The power of myth.* Doubleday: New York). This is in direct contrast to the popular sense of myth, which implies that something is false because it has no basis in fact or observation. For example, on websites such as the UK Royal Society and the Sierra Club of Canada, which expose some 'myths' of climate change, 'myths' are synonymous with 'misleading arguments'; i.e. scientific errors in the representation of climate change. That this popular meaning of 'myth' (= falsehood) is so pervasive itself reveals how dominant a purely materialist reading of reality has become.

climate change as an idea to be mobilised to fulfil a variety of tasks, perhaps we can see what climate change can do for us rather than what we seek to do, despairingly, for (or to) climate.

The idea of climate change is sufficiently plastic that it can be deployed across many of our human undertakings. We saw this in the opening to this chapter when surveying the diverse projects that array themselves in the language of climate change: the creative arts, rethinking intellectual property, securing energy, urban planning, poverty alleviation, gender equality, genetically modified crops. We can also mould the idea of climate change, knowingly or not, to serve many of our psychological, ethical and spiritual needs. This section suggests four ways in which our thoughts, discourses and feelings about climate change become loaded with deeper sets of assumptions about the worlds around us – and the worlds behind us and ahead of us – and our relationship with them. This is one sense in which climate change (lower case – physical transformation) becomes Climate Change (upper case – carrier of ideology; see Box 10.1).

I describe these ways of talking about climate change as 'myths', not in the popular sense of implying falsehoods, but in the specific anthropological sense of stories that 'embody fundamental truths underlying our assumptions about everyday or scientific reality'.[18] Myths in this non-pejorative sense become powerful shared narratives which may bind together otherwise quite different perspectives and people. The mystical, psychological, sociological and pedagogical functions of myths are more important than their 'scientific' truth or falsehood. Indeed, myths in this anthropological sense transcend the scientific categories of 'true' and 'false'.

The four myths of climate change I suggest here are rooted in our human instincts for nostalgia, fear, pride and justice. I am not

[18] p. 160 in Thompson, M. and Rayner, S. (1998) Risk and governance. Part I: The discourses of climate change. *Government and Opposition* 33(2), 139–66. See also Campbell, J. and Moyers, B. (1988) *The power of myth*. Doubleday; New York.

suggesting that these are the only myths that can be constructed around climate change, nor that the instincts attached to them necessarily reveal universal human values.[19] But I do believe they capture some of our most enduring psychological instincts as human beings. I attach metaphors – or avatars – to each myth to emphasise the constructed nature of these storylines and to offer a more evocative way of labelling them. Coming from a Western European culture with a heritage of Judaeo-Christian mythology, I adopt Biblical metaphors for these climate change myths: respectively, lamenting Eden, presaging Apocalypse, constructing Babel and celebrating Jubilee. Equivalent metaphors attached to the instincts of nostalgia, fear, pride and justice might also be found in other traditions such as Hindu, Norse and Chinese mythologies.

Lamenting Eden

This myth about climate change is related to the reading of climate we introduced in Chapter 1 under the heading 'The wildness of Nature'. I characterise this myth here using the Biblical image of a lost Eden; a picture which captures the idea of loss, lament and a yearning for restoration. As stated in the book of Genesis, 'So the Lord God banished him from the Garden of Eden to work the ground from which he had been taken. After he drove the man out, he placed on the east side of the Garden of Eden cherubim and a flaming sword flashing back and forth to guard the way to the tree of life.'[20]

In this lament, climate is viewed as a symbol of the natural or the wild, a manifestation of Nature that is pure and pristine and (should

---

[19] Over the last fifteen years, psychologist Shalom Schwartz has developed empirically the idea of universal human values. He argues that ten basic values can be recognised in cultures around the world, several of which map onto the four instincts I use here to develop the myths of climate change. See Schwartz, S. H., Melech, G., Lehmann, A., Burgess, S., Harris, M. and Owens, V. (2001) Extending the cross-cultural validity of the theory of basic human values with a different method of measurement. *Journal of Cross-Cultural Psychology* 32(5), 519–42.

[20] Genesis chapter 3, verses 23–4.

be) beyond the reach of humans. Climate becomes something that is fragile and needs to be protected or 'saved', goals which have fuelled the Romantic, wilderness and environmental movements of the Western Enlightenment over two centuries or more. A change to global climate brought about (inadvertently) by human action becomes a threat to this last remaining remnant of the wild. This is an idea that has been developed in part by the British sociologist Steve Yearley.[21] His suggestion is that we are concerned about climate change not so much because of any substantive diminution of human or non-human welfare that might ensue, but because of the strong element of symbolism involved. We are concerned about anthropogenic climate change because our climate has come to symbolise the last stronghold of Nature, untainted by Man.

When, where and why Nature first became a separate category in the human imagination can be argued over, as indeed can whether an independent category of wildness exists in any substantive sense.[22] Many anthropologists and environmental historians suggest that the idea of Nature *as* a separate category – distinct from Culture and therefore something that can be objectively studied by humans and hence physically 'damaged' by us – is an idea originating only in the Western Enlightenment. Such a position has led those in modern, rationalistic societies to distance themselves from Nature, a position that simultaneously, yet paradoxically, opens up both destructive and romantic tendencies. However one understands environmental and cultural history, there is no denying that the idea of wildness – Nature *as* separate – has been a persistent mode of discourse in Western rationalist cultures over recent centuries. It is an interpretation of

---

[21] Yearley, S. (2006) How many 'ends' of nature: making sociological and phenomenological sense of the end of nature. *Nature and Culture* 1(1), 10–21.

[22] See Cronon, W. B. (1996) The trouble with wilderness: or, getting back to the wrong nature, in Cronon, W. B. (ed.), *Uncommon ground: rethinking the human place in nature.* W. W. Norton & Co.: New York, pp. 69–90. This essay caused considerable uneasiness, even anger, within the American wilderness and environmental movements when it was published.

the world that finds rarer expression in traditional pre-modern or non-Western societies. In these settings, Nature and Culture are more consistently viewed as mutually embedded categories – there is no Nature unless interpreted by Culture and no Culture disembodied from Nature.

This climate change myth – what I call the lament for Eden – contends that by changing the climate, by losing wildness in one of the last 'untouched' places, humans believe they are diminishing not just themselves, but also something beyond themselves. We are the poorer for it – and maybe the gods are also. This mythic position emphasises the symbolic over the substantive and is a lament which underpins the deep ecology movement and which surfaces in some forms of eco-theology. It has also, I suggest, seeped more widely into mainstream environmentalism and perhaps lies hidden inside even broader climate change discourses across liberal Western societies.[23] Thus we find ecological economist Paul Baer challenging the World Bank economists by rhetorically asking what an ice sheet is worth, and the polar bear – that hackneyed icon of climate change – ends up not just worrying about its own survival but is made to carry a huge additional weight on its shoulders, the weight of human nostalgia. American photographer Camille Seaman's haunting exhibition in 2008, *The Last Iceberg*,[24] also played to this lament.

---

[23] For example, environmental historian Richard Grove writes about European colonial attitudes to tropical environments in the eighteenth and nineteenth centuries in terms of 'conserving Eden'. See Chapter 2 in Grove, R.H. (1998) *Ecology, climate and empire: the Indian legacy in global environmental history, 1400–1940*. Oxford University Press: Delhi, India.

[24] '*The Last Iceberg* chronicles just a handful of the many thousands of icebergs that are currently headed to their end. I approach the images of icebergs as portraits of individuals, much like family photos of my ancestors. I seek a moment in their life in which they convey their unique personality, some connection to our own experience and a glimpse of their soul which endures.' See www.camilleseaman. com/Artist.asp?ArtistID=3258&Akey=WX679BJN [accessed 28 May 2008].

Presaging Apocalypse

Environmental discourses have long been clothed in the language of Apocalypse.[25] Over ten years ago, literary analysts Jimmie Killingworth and Jacquie Palmer[26] traced part of this genealogy from the appearance of Rachel Carson's seminal book *Silent Spring* in 1962 to what they identified then, in 1996, as the new Apocalypse of global warming. Along the way they passed through Paul Ehrlich's 1968 *The Population Bomb* and the Club of Rome's 1972 *Limits to Growth*. Since they wrote this account (and, arguably, especially since 9/11), I suggest that the myth of Apocalyptic climate change has become even more dominant, particularly in Western European and North American discourses of climate change.

The linguistic repertoire of the Apocalypse draws upon categories such as 'impending disaster', 'approaching tipping points', 'species wiped out', 'billions of humans at risk of devastation, if not death'. As we saw in Chapter 6: *The Things We Fear*, there is an almost endless supply of headlines in print and on screen that are so phrased. We see evidence of this linguistic trope not just in the media and from environmental campaigners, but also in the words of civil servants such as Sir David King, the UK Government's then Chief Scientific Adviser – 'I believe that climate change is a bigger threat than global terrorism'[27] – and of some scientists such as James Lovelock: 'We [humans] are now so abusing the Earth that it may rise and move back to the hot state it was in 55 million years ago and if it does, most of us

---

[25] I use the word 'apocalypse' here in its popular sense, meaning destruction, rather than in its original Greek – and Biblical – form, meaning simply disclosure or revelation.

[26] Killingsworth, M. J. and Palmer, J. S. (1996) Millennial ecology: the apocalyptic narrative from 'Silent Spring' to global warming, in Herndl, C. G. and Brown, S. C. (eds), *Green culture: environmental rhetoric in contemporary America*. University of Wisconsin Press: Madison, WI, pp. 21–45.

[27] p. 176 in King, D. A. (2004) Climate change science: adapt, mitigate or ignore? *Science* 303, 176–7.

and our descendants will die.'[28] Visual imagery too becomes deployed in support of this myth of Apocalypse, the calving of ice from the Greenland Ice Sheet and maps of coastlines submerged under 7 metres of sea level rise being two popular visual cues. A separate category of climate change is invented – 'catastrophic climate change' as distinct from 'climate change'.

So why should climate change be portrayed in this way? I suggest at least three reasons for the myth of a climate Apocalypse that we have seen established in recent years. First is the enduring human fear of the future which fuels these descriptions of a physical climate system on the point of collapse. It sits easily with the view of Nature as fragile and ephemeral, as we saw in Chapter 6 (Box 6.1). Second, this myth of Apocalypse also draws strength from the new paradigm of Earth system science with its ideas of complexity, thresholds and tipping elements. Working within this paradigm, Earth system models are able to find an increasing number of 'choke points' in their modelled worlds where non-linear changes in virtual climate function can occur. The movie *The Day After Tomorrow*, which we looked at in Chapter 7: *The Communication of Risk*, developed its wildly exaggerated storyline around one such tipping element – the collapse of the thermohaline ocean circulation in the North Atlantic.

A third reason for the salience of this myth – this way of talking about climate change – is the frustration experienced by some campaigners and policy advocates due to the failure of international measures and agreements to start slowing the growth in carbon emissions. The instinctive reaction of these commentators to such perceived failure is to turn the dial on the amplifier a notch higher and proclaim that the risks of climate change are greater than first thought. This latter mentality was revealed by American climate change expert Stephen Schneider, when being interviewed back in

---

[28] p. 1 in Lovelock, J. (2006) *The revenge of Gaia: why the earth is fighting back – and how we can still save humanity.* Penguin: London.

October 1988: 'We need to get some broad-based support, to capture the public's imagination ... so we have to offer up scary scenarios, make simplified, dramatic statements and make little mention of any doubts ... each of us has to decide what the right balance is between being effective and being honest.'[29]

The myth of climate change as Apocalypse therefore appeals to our instinct of fear about the unknown future, but it also acts as a call to arms. The purpose of environmental crisis rhetoric has always been to change the future rather than to predict it. Paul Ehrlich, for example, claimed that by painting in the late 1960s a scenario of a dysfunctional and Malthusian world owing to unfettered population growth, he in fact contributed to a slowing down of population growth, thereby averting the very scenario he foresaw. A similar normative goal may be being served by portrayals of climatic Apocalypse. This normative role for climate apocalyptic myth-making enlists powerful scientific predictions in its cause, drawing legitimacy from IPCC science and, in this sense, may therefore be different from earlier 1960s environmental apocalyptic rhetoric. Yet this myth is easily moulded into different forms and can lend support to a number of the different policy discourses we examined in Chapter 8: *The Challenges of Development*. Radical ecology, ecological modernisation, social activism and neoliberal conservatism are all projects which can claim ascendancy from the association of climate change with fear of the future.

What does the myth of climatic Apocalypse do to audiences around the world? It undoubtedly lends a sense of danger, fear and urgency to discourses around climate change captured in claims such as we only have ten years to 'reduce emissions' or to 'save the climate'. Yet the

[29] Schneider, S. H. (1988) Interview in *Discover* magazine, New York, October. Schneider appears here to be sacrificing 'honesty' for 'effectiveness' by eliminating doubt and uncertainty. When later reflecting on this 1989 interview – *Times Higher Education*, 21 August 2008, p. 43 – Schneider regrets having given this impression and suggests that there are ways of being both honest and effective through the use of metaphors that 'convey both urgency and uncertainty'.

counterintuitive outcome of such language is that it frequently leads to disempowerment, apathy and scepticism among its audience. Many studies – and not only in relation to climate change – have shown that promoting fear is often an ineffective, even counterproductive, way of inducing behavioural change. Communication expert Susi Moser has concluded that 'numerous studies show that … fear may change attitudes … but not necessarily increase active engagement or behaviour change'.[30] If such change really is what deployment of this myth is seeking to accomplish, then it may be self-defeating.

Constructing Babel

My third mythic interpretation of climate change I call constructing Babel. This myth is not necessarily orthogonal to either of the previous two, and indeed it gains some of its legitimacy from narrating climate change as Apocalypse. As the Genesis myth of Babel relates, a confident and independent humanity, re-populated after the traumas of the Flood, claimed, 'Come, let us build ourselves a city, with a tower that reaches to the heavens, so that we might make a name for ourselves and not be scattered over the face of the Earth.'[31] This aspiration towards god-like status, acclaim and personal glory exemplifies the Greek idea of hubris – excessive self-confidence and the desire to dominate.

The eminent nuclear scientist and mathematician John von Neumann foresaw this myth of Babel in relation to climate control more than half a century ago, writing in 1955 that 'intervention in atmospheric and climatic matters will come in a few decades and will unfold on a scale difficult to imagine at present … what power over our environment, over all Nature, is implied!'[32] This myth also mobilises

---

[30] p. 70 in Moser, S. (2007) More bad news: the risk of neglecting emotional responses to climate change information, in Moser, S. and Dilling, L. (eds), *Creating a climate for change: communicating climate change and facilitating social change*. Cambridge University Press, pp. 64–81.

[31] Genesis chapter 11, verse 4.

[32] von Neumann, J. (1955) Can we survive technology? *Fortune*, June issue. Reproduced in *Population and Development Review* (1986) 12(1), 117–26.

the idea of climate utopias and our ability to engineer them for human benefit, an idea explored further in Box 10.2. More recently, American sociologist John Lie, while qualifying von Neumann's confident claims about ultimate climate control, alludes to the same mindset when writing about climate change and politics. 'Even if God-like control is unattainable – and the very idea of the conquest of nature is hubris – a deeply optimistic strain in the scientific mindset envisions nature as increasingly colonised and controlled by human conceptual inventions and technological interventions.'[33]

Box 10.2: Climate Utopias

Few of us consider that our climate is optimal, or at least very few of us consider that it is always optimal. It is too unpredictable (or too predictable), too wet (or too dry), too cloudy (or too hot). This state of emotional discontent with our climate also afflicted our ancestors. In their imagination of an earthly paradise, writers in classical antiquity removed extreme temperatures, strong winds and overcast skies from their utopian weather, while the study of utopian novels has found that the genre usually presents the climate 'either as an equable given or something totally under man's [sic] control'.[34]

This has not prevented us from attaching 'optimal' labels to certain types of climates, either climates of particular places or climates of particular eras. We saw in Chapter 1: *The Social Meanings of Climate* how Ellsworth Huntington and geographers of similar ilk at the beginning of the twentieth century found empirical

---

[33] p. 234 in Lie, J. (2007) Global climate change and the politics of disaster. *Sustainability Science* 2, 233–6.

[34] p. 210 in Porter, P. W. and Lukermann, F. E. (1976) The geography of utopia, in Lowenthal, D. and Bowden, M. J. (eds), *Geographies of the mind: essays in historical geosophy in honour of John Kirtland Wright*. Oxford University Press: New York.

'proof' that humans worked best and thought most creatively in the 'optimal' climates of southern England and the north-eastern states of America. Historians later labelled the global climate of the early Holocene era, between 5,000 and 8,000 years ago, as 'optimal' and described the climates of the North Atlantic region between the eleventh and fourteenth centuries as representing a secondary (Medieval) 'climatic optimum'.

This uneasy relationship with our climate, together with our ability to imagine climate utopias, has frequently turned our thoughts to ideas of climate control and engineering. From Charles Fourier, the French visionary who laid out the first blueprint for a human-engineered climate in the 1820s, to the new geo-engineers of the twenty-first century such as Paul Crutzen and James Lovelock, we have dreamed and schemed of finding ourselves masters of the world's climates. As with the lament for Eden, this utopian and masterful tendency also has echoes of an Edenic project. Rather than returning to an eternally natural paradise and preserving it against the ravages of humanity, the techno-centric deployment of this myth is that Eden can be perpetually remade through the ingenuity and managerial capacities of humanity. Perfected climates, it seems, can only be found by expunging humans on the one hand, or by mastering climate on the other.

But whose climate are we seeking to optimise? Whose climate utopia are we seeking to engineer? Was the global climate of the early Holocene one to which we wish to return and what about the Medieval optimum of the twelfth century in Europe or the pre-industrial global climate of the nineteenth century? Constructing a global thermostat – whether through mirrors in space or through injections of sulphur dioxide into the stratosphere – is a Procrustean option for delivering climate utopia to the masses. And which masses? Those who speak loudest, those who pay the most, or those who are condescendingly judged to be most in need of a dose of (our) utopian climate?

Our descendants may not thank us for the legacy of an atmosphere over-supplied with greenhouse gases, but neither may they thank us for handing them the gift of a global thermostat over the control of which their future wars may be fought. Maybe better than seeking climate utopia through exercising our power over Nature would be to learn from the Oneida Community of nineteenth-century Connecticut in the USA. The 'Perfectionists' as they called themselves, set about adapting themselves and their way of life to the 'weather God gave them'. The Oneidans' view of the weather was 'something whose meaning and significance depended entirely on the ways of life carried on within it'.[35] This was an internal climate utopia shaped by the imaginative and cultural practices of community, rather than an external one controlled and managed by a technocratic elite.

I suggest that this confident belief in the human ability to control Nature is a dominant, if often subliminal, attribute of the international diplomacy that engages climate change. The challenges raised by climate change are seen in essentially modernist, techno-managerial terms. As with acidic air, chemical toxins and ozone depletion, a risk is identified and the apparatus of the State – or in the case of climate change the apparatus of many States – is mobilised to mitigate the risk. Perhaps drawing emotional power from the domineering language of climate change as Apocalypse, the management of climate change becomes the latest project over which human governance, control and mastery is demanded. Thus, leading climate scientist Jim

[35] p. 595 in Meyer, W.B. (2002) The Perfectionists and the weather: the Oneida Community's quest for meteorological utopia 1848–1879. *Environmental History* 7(4), 589–610.

Hansen's recent appeal to the 'loss' of control as a reason for concern. 'An important point is that the non-linear response could easily run out of control [whose control does he mean?], because of positive feedbacks and system inertias.'[36]

This myth of climate mastery and control has an interesting variant, the sub-myth of geo-engineering. This narrative – drawing power from the new modelling claims of the Earth system scientists – argues that exactly *because* we are unlikely to realise our Tower of Babel using the conventional instruments of diplomacy, trade and regulation, we need a new form of intervention to bring a runaway climate under direct human management. Scientists James Lovelock and Chris Rapley capture this mood perfectly when they reveal their belief that '[amplifying] feedbacks, the inertia of the Earth system – and [the inertia] of our response – make it doubtful that any of the well-intentioned technical or social schemes for carbon dieting will restore the status quo'.[37] They then propose a large-scale scheme for sucking up cold, nutrient-rich waters from the deep ocean to the surface, where they can fertilise algae which will feed on carbon dioxide in the atmosphere. 'If we can't heal the planet directly, we may be able to help the planet heal itself.'

There is a deep irony displayed in this sub-myth. The inadvertent side-effects of carbon-fuelled economic growth – the 'great geophysical experiment' with the planet first described by Roger Revelle in 1957 – are charged with destabilising our naturally regulated climate. All efforts to reign in the damage using conventional human control systems are deemed to be failing. What is therefore proposed is a new, but now deliberate, great geophysical experiment with the planet. The only difference between this new purposeful experiment and our

---

[36] p. 4 in Hansen, J. E. (2007) Scientific reticence and sea level rise. *Environmental Research Letters* 2 doi:10.1088/1748–9326/2/2/024002.

[37] p. 403 in Lovelock, J. and Rapley, C. (2007) Ocean pipes could help the ocean to cure itself. *Nature* 449, 403.

ongoing inadvertent one is that we now have the 'wisdom' of Earth system models to guide us.

Whether one believes in the finance offices of the World Bank, the efficacy of the traditional nation-state, or the scientific high priests of Gaia, believing that we can 'make a name for ourselves' by mastering and stabilising global climate requires an inordinate degree of faith.

Celebrating Jubilee

The fourth myth of climate change I offer relates to the ways in which we think and talk about climate change using the language of morality and ethics. We examined some of the evidence for this in Chapter 5: *The Things we Believe* – the inescapability of adopting ethical positions when we diagnose responsibility for climate change and propose responses, or the way in which the language of popular discourse about climate change and responsibility echoes the theological language of sin and repentance. Here I suggest that our instinct for justice fuels this powerful myth about climate change. For some, the desire for justice is synonymous with the entire meaning of climate change. I attach to this myth the metaphor of Jubilee – the idea of justice, freedom and celebration; an idea embedded in the Jewish Torah that, every fifty years, soil, slaves and debtors should be liberated from their oppression. 'In this way you shall set the fiftieth year apart and proclaim freedom to all the inhabitants of the land.'[38] The occasion was to be marked by a year-long celebration.

For those in social and/or environmental justice movements, climate change is not primarily a substantive, material problem, nor simply – as in the lament for Eden – a symbolic one. Climate change is an idea around which their concerns for social and environmental justice can be mobilised. Indeed, a new category of justice – climate justice – is demanded, and one that attaches itself easily to other long-standing global justice concerns: 'By and large, the framing of "climate justice"

---

[38] Leviticus chapter 25, verse 10.

reflects the same social and economic rights perspectives voiced by global movements on debt, trade and globalisation.'[39] A former environment minister in Canada's Liberal Government of the late 1990s clearly revealed the underlying commitment to this mythic position: 'No matter if the science of global warming is all phony ... climate change provides the greatest opportunity to bring about justice and equality in the world.'[40]

This instinct for justice, and the recognition that climate change demands an engagement with this instinct, is not just to be found within radical social movements. Ideas of justice and equity are threaded through political debates and international negotiations about climate change policies, whether in the context of procedural decision making, emissions reduction burden-sharing, or vulnerability, adaptation and compensation. And some would argue that, with respect to ordinary citizens, this mythic position of climate change as social justice has instinctive, mobilising power. It appeals to certain intrinsic values held by many – values of fairness and justice around which popular (Western) campaigns such as Make Poverty History have been able to mobilise.

By seeing climate change as a celebration of Jubilee – at least as an opportunity to create new potency behind movements for social and environmental justice – this myth offers hope as an antidote to the presaging of Apocalypse. Thus ethicist James Garvey concludes his book *The Ethics of Climate Change: Right and wrong in a warming world* with the observation: 'Many ... say something remarkable, something uplifting and hopeful. They say that climate change offers humanity the chance to do the right thing.'[41] And campaigner Alastair

---

[39] p. 102 in Pettit, J. (2004) Climate justice: a new social movement for atmospheric rights. *IDS Bulletin* 35(3), 102–6.

[40] Christine Stewart, speaking in 1998. Quoted in *Canada Free Press*, 5 February 2007.

[41] p. 155 in Garvey, J. (2008) *The ethics of climate change: right and wrong in a warming world*. Continuum International Publishing: London and New York.

McIntosh, in his book entitled *Hell and High Water: Climate change, hope and the human condition*,[42] further elaborates the presence of hope within this mythic position on climate change.

10.4 The Cultural Messages of Climate Change

We saw at the beginning of the book how our sensual experiences and scientific depictions of physical climates have historically been inextricably entangled with meanings reflecting broad cultural and ideological movements. Now, at the end of the book, I argue that it is climate change and not climate *per se* that has become the new locus of such entanglements. We won't understand climate change by focusing only on its physicality. We need to understand the ways in which we talk about climate change, the variety of myths we construct about climate change and through which we reveal to ourselves what climate change means to us.

The Enlightenment project to objectivise climate through standardised measurement and quantification has led us moderns to see climate change as a physical transformation to be predicted and managed, if not mastered. Disagreements about this project too easily revolve around disputed science, contentious economics and precarious diplomacy. Having allowed science to strip climate of its cultural anchors and ideological meanings – part of what sociologist Bruno Latour calls the 'purification' of knowledge – we have been slow to recognise that understanding climate change and responding to it demands a re-engagement with the deeper and more intimate meanings of climate that have been lost. Science may be solving the mysteries of climate, but we have lost – or chosen to ignore – what climate means to us. These meanings lie beyond the reach of science, economics and politics. This is an observation reflected in a more general

---

[42] McIntosh, A. (2008) *Hell and high water: climate change, hope and the human condition*. Birlinn: Edinburgh.

context by English playwright Tom Stoppard in his 1993 play *Arcadia*, where his character Septimus remarks, 'When we have found all the mysteries and lost all the meaning, we will be alone, on an empty shore.'

The international Alliance of Religions and Conservation has also understood this irony. Their seven-year climate change programme statement proclaims that 'the emphasis on consumption, economics and policy usually fails to engage people at any deep level because it does not address the narrative, the mythological, the metaphorical or the existence of memories of past disasters and the way out'. And in a world where the globalising powers of capital, trade and consumption separate us from the local stories that give meaning not just to climate, but also frequently to our lives, this perspective also offers a way of reconnecting ourselves with climate through the telling of stories.

> Without … these areas [of narrative, myth and metaphor], policies will have very few real roots … the climate change 'activist' world and indeed the environmental world has all too often sought refuges in random use of apocalyptic imagery without seeking to harness the power of narrative. Without narrative, few people are ever moved to change or adapt. [43]

Climate change, then, can serve a different purpose from the one our modern mind instinctively thinks of. Just as climate cannot ultimately be cleaved from our cultural and personal imaginations, so neither can climate change. We are creating ways to give new meaning to climate change as it circulates through our imaginative and social worlds, even as we embrace the 'purified' physical form of climate change offered us by the universalising messages of the IPCC.

Climate change is doing deeper cultural work for us, moving us away from a modern reading of climate and either back to a pre-modern or forward to a post-modern reading. We are finding new ways

---

[43] Alliance of Religions and Conservation (2007) *ARC/UNDP Programme Statement on Climate Change*, 7 December. See www.arcworld.org/news.asp?pageID=207 [accessed 21 July 2008].

of reinserting culture, ideology and meaning into our experiences of climate, both physically present climates and virtually future climates. We may no longer see climate as the domain of local gods or a reflection of local (im)moral behaviours, but we can use the idea of global climate change to tell ourselves new stories about our globalised gods and about the consequences of our collective behaviours. We can use the idea of climate change to move beyond the separated categories of the physical and the cultural, beyond the framing of climate change that uses the language of problem and solution. Climate change can help us bring the physical and the cultural, the material and spiritual, into a new realignment.

Climate change thus becomes a mirror into which we can look and see exposed both our individual selves and our collective societies. We can use the stories we tell about climate change – the myths we construct – to rethink the ways in which we connect our cultural, spiritual and material pursuits. It does not necessarily make it any easier for us to agree about climate change, although it suggests that we are perhaps seeking agreement about climate change in the wrong places and for the wrong reasons.

The four myths I introduced in the previous section suggest ways in which we can offer visibility to this idea. These myths – these ways of ordering reality – don't map directly onto the four ways of living which Cultural Theory introduced us to earlier in the book (see Chapter 6: *The Things We Fear*). But each myth no doubt resonates differently with these larger cultural world-views. Thus egalitarians are perhaps more likely to engage with the myths of Eden, Jubilee and Apocalypse, hierarchists perhaps with the myths of Apocalypse and Babel, and individualists perhaps with the myth of Babel.

All four myths engage seriously with the physical transformations of our climate which science is revealing to us. There are observable and prospective changes in physical climates around the world and humans are active agents in these changes. Yet each myth goes well beyond the mere physicality of climate change and ends up telling richer stories about what these changes signify. It is stories such as these – embodiments of 'fundamental truths about our assumptions of reality' – that we need to re-create in our world. Climate change offers great story-telling potential. The four myths I have offered should not be judged as either right or wrong. They should be recognised as stories about climate change; as mirrors that reveal important truths about the human condition.

The myth of Eden, born of nostalgia, tells us of our desire – our yearning even – to return to some simpler era when the domains of Nature and the definitions of the natural seemed clearer and unambiguous. In the lament for Eden we are telling ourselves that in the active shaping of our climate we have appropriated god-like powers, yet they are powers we are uncomfortable with.

The myth of Apocalypse, born of fear, tells us of our worry about the future. No longer is it the capricious climate gods whom we need to respect and appease to ensure our survival, but it is ourselves and our voracious appetite for material consumption that we now fear. We have lost the sense of transcendent mystery and gratitude that once offered restraint to this appetite and we do not know where to re-locate it.

The myth of Babel, born of pride, tells us of our desire for mastery and control. We have always exercised this instinct over our fellow human beings and our fellow creatures. Climate change now offers us a global domain over which we can create new instruments and institutions of control, but it is a domain and a project in which we will never reach our hubristic goal of the ideal state.

And the myth of Jubilee, born of justice, tells us of the inescapable call for humans to respond to injustice. Climate change opens out for us new ways of understanding the wilful and structural causes of inequality and injustice in the world, and challenges our instinct to respond. But we see, too, in climate change how this instinct for justice clashes with the structures that hem us in, and how it reveals the limits of our individual moral agency.

10.5 Beyond Climate Change

The book opened in Chapter 1 with a survey of the meanings of climate. I identified four themes which have acted as a framework for the ideas of climate and climate change around which this book has been written. These themes were: that the idea of climate has both physical and cultural dimensions; that the idea of climate has consequently been used by societies and movements to carry and promote different ideological projects; that climates change over time, both physical climates and the ideas that climate carries for us; and, finally, that the stories we tell about the relationship between climate change and human civilisation also change over time.

In this concluding chapter I have used these underlying themes to help look beyond climate change, at least to look beyond the mere physicality of climate change as an environmental problem to be solved. I have argued that because climate change is a 'wicked problem' it does not lend itself to a solution; either elegant solutions or clumsy ones. If we pursue the route of seeking ever larger and grander solutions to climate change we will continue to end up frustrated and disillusioned: global deals will be stymied, science and economics will remain battlegrounds for rearguard actions, global emissions will continue to rise, vulnerabilities to climate risks will remain. And we will end up unleashing ever more reactionary and dangerous interventions in our despairing search for a solution to our wicked problem: the colonisation of agricultural land with energy crops, the colonisation of space with mirrors, the colonisation of the human spirit with authoritarian government.

Instead, I have presented climate change as an idea, an imaginative resource, which can be made to do work for us. I have put forward four myths – stories that embody truths about how we assume reality to be – about climate change, each of which captures different aspects of ways in which humans see themselves, relate to each other and seek to live on the Earth. Myths are often found embedded in religious ways of looking at the world. Indeed, many secular myths have religious foundations. In his book *Faith in Nature: Environmentalism as religious quest*, environmental historian Thomas Dunlap offers the view that religions are less about solving problems than about confronting them. Just as Dunlap argues that, by adopting this religious perspective, environmentalism can help us confront the human situation in our world; as a specific case, so too can climate change. The four myths offered earlier in the chapter are attempts to do just this by holding up mirrors in which we can see ourselves reflected – reflections which reveal the contradictions and limitations that make us human.

Thus, we talk about climate change using the language of lament and nostalgia, revealing our desire to return to some simpler, more innocent, era. We are uneasy with the unsought powers to change global climate we now have. We talk about climate change using the language of fear and apocalypse, revealing our endemic worry about the future. We have lost the sense of transcendent mystery and gratitude that once offered us conduits for defusing these fears. We talk about climate change using the language of pride and control, revealing our desire for dominance and mastery. Climate change now offers us a global domain for such mastery, but we lack the wisdom and humility to exercise it. And we talk about climate change using the language of justice and equity, revealing the inescapable call for humans to redress revealed injustices. But climate change also reveals the limits of our individual moral agency.

What Dunlap calls a 'religious perspective' is a point of view that 'helps sort out the arguments, beliefs and actions, clears away mental underbrush, and allows people to explore issues now hidden because not seen'.[44] Therein lies the value in offering these mythical stories that underpin our discourses about climate change. If we continue to

---

[44] p. 171 in Dunlap, T. R. (2004) *Faith in nature: environmentalism as religious quest.* University of Washington Press: Seattle, WA.

talk about climate change as an environmental problem to be solved, if we continue to understand the climate system as something to be mastered and controlled, then we have missed the main lessons of climate change. If climate means to us only the measurable and physical dimensions of our life on Earth then we will always be at war with climate. Our climates will forever be offering us something different from what we want.

Rather than placing ourselves in a 'fight against climate change', we need a more constructive and imaginative engagement with the idea of climate change. Clarence Glacken, learning from his 1967 survey of Western environmental thought, recognised the essence of this challenge when he wrote a few years later:

> If the history of thought teaches us anything about culture and environment, it is the importance of the conceptions which people have of both – whether these conceptions are religious, philosophical, scientific or utilitarian. It is a poor time in history to have a narrowness of vision and to continue to limit values to ideas of struggles and antithesis. The diversity of world cultures demands much more of us in understanding and imagination.[45]

The physical transformation of our climates now under way shows both the extent of our inadvertent and unwanted agency, but also the limits of our science-saturated and spiritually impoverished wisdom. Humility thus becomes a virtue. We need the 'modes of knowing that ... supplement science with those aspects of the human condition that science cannot easily illuminate'[46] referred to by Sheila Jasanoff in her essay on the technologies of humility.

The idea of climate change should be used to rethink and renegotiate our wider social goals about how and why we live on this planet. We need to harness climate change to give new expression to some of

[45] Glacken, C. (1970) Man against Nature: an outmoded concept, in Helrich, H. W. (ed.), *The environmental crisis: Man's struggle to live with himself.* Yale University Press: New Haven, CT.
[46] p. 33 in Jasanoff, S. (2007) Technologies of humility. *Nature* 450, 33.

the irreducible and intrinsic human values that are too easily crowded out – our desires for personal growth and self-determination, for creative experimentation, for relationship and for community. In this way, climate change can be assimilated into our future. If we harness the full array of human sciences, artistic and spiritual endeavours, and our civic and political pursuits we *can* reconcile climate change with our human and social evolution, with our instinct for justice and with our endurance on this planet. Human ecologist Jesse Ausubel recently wrote along these lines in drawing lessons from Europe's medieval past:

> Great sins can elicit great cathedrals. In fact, the people of medieval Europe were not more evil than those of other times and places, but they channelled their guilt to glorious enduring expression. Let us similarly channel the diffuse anxiety that is environmentalism into immense achievement … we can avoid screams while lessening worries about climate change.[47]

The function of climate change I suggest, then, is not as a lower-case environmental phenomenon to be solved. Solving climate change should not be the focus of our efforts any more than we should be 'solving' the idea of human rights or liberal democracy. It really is not about stopping climate chaos. Instead, we need to see how we can use the idea of climate change – the matrix of ecological functions, power relationships, cultural discourses and material flows that climate change reveals – to rethink how we take forward our political, social, economic and personal projects over the decades to come.

We should use climate change both as a magnifying glass and as a mirror. As a magnifier, climate change allows us to conduct examinations – both more forensic and more honest than we have been used to – of each of our human projects: whether they be projects of personal well-being, self-determination, liberated or localised trade,

---

[47] Ausubel, J. (2001) Some ways to lessen worries about climate change. *Electricity Journal* (Jan/Feb), 24–31.

poverty reduction, community-building, demographic management, or social and psychological health. Climate change demands that we focus on the long-term implications of short-term choices, that we recognise the global reach of our actions, and that we are alert both to material realities and to cultural values. And as a mirror, climate change teaches us to attend more closely to what we really want to achieve for ourselves and for humanity. Yet the myths I have labelled Eden, Apocalypse, Babel and Jubilee reflect back to us truths about the human condition – instincts – that are both comforting and disturbing. They suggest that even were we to know what we wanted – personal affluence, social justice or mere survival – we are limited in our abilities to acquire or deliver it.

As a resource of the imagination, the idea of climate change can be deployed around our geographical, social and virtual worlds in creative ways. The idea of climate change can stimulate new thinking about technology. It can inspire new artistic creations in visual, written and dramatised media. It can invigorate efforts to protect our citizens from the hazards of climate. The idea of climate change can provoke new ethical and theological thinking about our relationship with the future. It can arouse new interest in how science and culture interrelate. It can galvanise new social movements to explore new ways of living in urban and rural settings. And the idea of climate change can touch each one of us as we reflect on the goals and values that matter to us. These are all creative applications of the idea of climate change, but they are applications that do not demand agreement. Indeed, they may be hindered by the search for agreement. They thrive in conditions of pluralism and hope, rather than in conditions of universalism and fear. Nor are they applications that will lead to stabilising climate – they will not 'solve' climate change.

The world's climates will keep on changing, with human influences on these physical properties of climate now inextricably entangled with those of Nature. Global climate is simply one new domain which reveals our embeddedness in Nature. But so too will the idea of climate change keep changing as we find new ways of using it to meet our needs. We will continue to create and tell new stories about climate change and mobilise these stories in support of our projects. Whereas a modernist reading of climate may once have regarded it as merely a physical boundary condition for human action, we now must come to terms with climate change operating as an overlying, but more fluid, imaginative condition of human existence.

If used wisely, this condition of climate change in which we are now embroiled may teach and empower us to embark on different projects from those that come easily to us. Even so, we will disagree. But let us at least recognise that the sources of our disagreement about climate change lie deep within us, in our values and in our sense of identity and purpose. They do not reside 'out there', a result of our inability to grasp knowingly some ultimate physical reality. And let us remember that our disagreements should, at best, always lead us to learn more about ourselves – our lament for the past, our fear of the future, our desire for control, and our instinct for justice. Our engagement with climate change and the disagreements that it spawns should always be a form of enlightenment.

FURTHER READING FOR CHAPTER 10

Glover, L. (2006) **Postmodern climate change.** Routledge: London.
Glover's book approaches the phenomenon of climate change and the social responses to it using the theories, tools and language of post-modernity. It offers a wide-ranging survey of the science and politics of climate change and of the social discourses and movements that have emerged in response to it.

Malone, E.L. (2009) **From debate to agreement: the climate change issue.** Earthscan: London.
This book explores how people argue about climate change, what underpins their positions, and whether these positions are mutable or ingrained. Elizabeth Malone draws upon over 100 published documents on climate change from a wide range of sources to offer an empirical analysis of different discourses, and suggests ways in which more general agreement might be found.

McIntosh, A. (2008) **Hell and high water: climate change, hope and the human condition.** Birlinn: Edinburgh.
McIntosh suggests that politics alone is not enough to tackle climate change. At root is our addictive consumer mentality and, in a journey through early texts that speak to climate change, McIntosh reveals the psychohistory of modern consumerism. If climate change is a spiritual problem, then its solutions must also be spiritual.

Simms, A. and Smith, J. (eds) (2008) **Do good lives have to cost the earth?** Constable: London.
This book offers a series of personal visions from twenty-one figures in British public life about how concerns about climate change are integrated into lifestyles and lifestyle choices. It is unusual in that it seeks to reveal how individuals integrate personal values with their wider sets of understandings about climate change and what it signifies.

Verweij, M. and Thompson, M. (eds) (2006) **Clumsy solutions for a complex world: governance, politics and plural perspectives.** Palgrave Macmillan: Basingstoke, UK, and New York.
Through a series of case studies looking at how various social and environmental ills have been tackled around the world, this book introduces the idea of clumsy solutions in contra-distinction to the prevalence of elegant failures. Drawing upon Cultural Theory, it argues that we have to allow multiple ways of living and viewing the world to co-exist, with all of their complex contradictions, if we are to make any headway in tackling the problems that beset us. Clumsy solutions based on plurality are the opposite of elegant optimisation based on dogma.

Bibliography

Adams, W. M. (2008) *Green development: environment and sustainability in the South* (3rd edn). Routledge: London.

Adger, W. N., Paavola, J., Huq, S. and Mace, R. J. (2006) *Equity and justice in adaptation to climate change*. MIT Press: Cambridge MA.

Adger, W. N., Lorenzoni, I. and O'Brien, K. (eds) (2009) *Adapting to climate change: thresholds, values, governance*. Cambridge University Press.

Agassiz, J.-L. (1840) *Études sur les Glaciers*. Neuchatel, Switzerland.

Agrawal, A. and Narain, S. (1991) *Global warming in an unequal world*. Centre for Science and the Environment: Delhi, India.

Aldy, J. E. and Stavins, R. N. (eds) (2007) *Architectures for agreement: addressing global climate change in the post-Kyoto world*. Cambridge University Press.

All Party Parliamentary Group on Population, Development and Reproductive Health (2007) *Return of the population growth factor: its impact on the Millennium Development Goals*. HMSO, London.

Allan, S., Adam, B. and Carter, C. (eds) (2000) *Environmental risks and the media*. Routledge: London.

Anon. (2006) As Earth warms, Congress listens. *Science* **314**, 29.

Arrhenius, S. (1896) On the influence of carbonic acid in the air upon the temperature of the ground. *London, Edinburgh and Dublin Philosophical Magazine and Journal of Science* **41**, 237–76.

Atkinson, D. (2008) *Renewing the face of the Earth: a theological and pastoral response to climate change*. Canterbury Press, London.

Atran, S., Axelrod, R. and Davis, R. (2007) Sacred barriers to conflict resolution. *Science* **317**, 1039–40.

Ausubel, J. (2001) Some ways to lessen worries about climate change. *Electricity Journal* (Jan/Feb), 24–31.

Bäckstrand, K. and Lövbrand, E. (2008) Climate governance beyond 2012: competing discourses of green governmentality, ecological modernisation and civic environmentalism. In: Pettenger, M. (ed.), *The social construction of climate change*. Ashgate: Aldershot, UK, pp. 123–41.

Bailey, I. (2007) Market environmentalism, new environmental policy instruments and climate policy in the UK and Germany. *Annals of the Association of American Geographers* **97**(3), 530–50.

Bailey, I. (2007) Neoliberalism, climate governance and the scalar politics of EU emissions trading. *Area* **39**(4), 431–42.

Banuri, T. and Weyant, J. (eds) (2001) Setting the stage: climate change and sustainable development. Chapter 1 in *Climate change 2001: mitigation*. Report of Working Group III to the Intergovernmental Panel on Climate Change. Cambridge University Press.

Barrett, S. (2008) *Climate change negotiations reconsidered*. Briefing Paper, Policy Network: London.

Beck, R. A. (1993) Viewpoint: climate, liberalism and intolerance. *Weather* **48**, 63–4.

Behrensmeyer, A. K. (2006) Climate change and human evolution. *Science* **311**, 476.

Benedick, R. (1998) *Ozone diplomacy: new directions in safeguarding the planet* (2nd edn). Harvard University Press: Cambridge, MA.

Berlin, I. (2002) *Liberty* (revised and expanded edition of *Four Essays On Liberty*, 1969). Oxford University Press: New York.

Boia, L. (2005) *The weather in the imagination* (trans. R. Leverdier). Reaktion Books: London.

British Council (2005) *Talking about climate change*. British Council: Manchester, UK.

Broecker, W. S. (1987) Unpleasant surprises in the greenhouse? *Nature* **328**, 123–6.

Broecker, W. S. (1999) What if the conveyor were to shut down? Reflections on a possible outcome of the great global experiment. *Geological Society of America Today* **9**(1), 1–7.

Brönnimann, S. (2002) Picturing climate change. *Climate Research* **22**, 87–95.

Brooks, C. E. P. (1949) *Climate through the ages: a study of the climatic factors and their variations* (2nd edn). Ernest Benn: London.

Brooks, N. (2006) Cultural responses to aridity in the Middle Holocene and increased social complexity. *Quaternary International* **151**, 29–49.

Brückner, E. (1890) *Klimaschwankungen seit 1700 nebst Bemerkungen über die Klimaschwankungen der Diluvialziet*. Wien and Olmütz.

Bryson, R. A. and Murray, T. J. (1977) *Climates of hunger: mankind and the world's changing weather*. University of Wisconsin Press: Madison, WI.

Buesseler, K. O., Doney, S. C., Karl, D. M., Boyd, P. W., Caldeira, K., Chai, F., Coale, K. H., de Baar, H. J. W., Falkowski, P. G., Johnson, K. S., Lampitt, R. S., Michaels, A. F., Naqvi, S. W. A., Smetacek, V., Takeda, S. and Watson, A. J. (2008) Ocean iron fertilization – moving forward in a sea of uncertainty. *Science* **319**, 161.

Buffon (1778) *Les époques de la Nature*. Paris.

Callendar, G. S. (1938) The artificial production of carbon dioxide and its influence on temperature. *Quarterly Journal of the Royal Meteorological Society* **64**(2), 223–40.

Callendar, G. S. (1939) The composition of the atmosphere through the ages. *Meteorological Magazine* **74**, 33–9.

Campbell, J. and Moyers, B. (1988) *The power of myth*. Doubleday: New York.

Campbell, M., Cleland, J., Ezeh, A. and Prata, N. (2007) Return of the population growth factor. *Science* **315**, 1501–2.

Carley, M. and Christie, I. (2001) *Managing sustainable development* (2nd edn). Earthscan: London.

Carson, R. (2000) *Silent spring*. Penguin Classics: London. First published in 1962.

Carvalho, A. (2007) Communicating global responsibility? Discourses on climate change and citizenship. *International Journal of Media and Cultural Studies* **3**(2), 180–3.

Carvalho, A. (2007) Ideological cultures and media discourses on scientific knowledge: re-reading news on climate change. *Public Understanding of Science* **16**, 223–43.

Carvalho, A. and Burgess, J. (2005) Cultural circuits of climate change in the UK broadsheet newspapers, 1985–2003. *Risk Analysis* **25**(6), 1457–70.

Chishti, S. (2003) *Fitra*: an Islamic model for humans and the environment. In: Foltz, R., Denny, F. and Bahaaruddin, A. (eds), *Islam and ecology: a bestowed trust*. Harvard University Press: Cambridge, MA, pp. 67–82.

Collier, P. (2007) *The bottom billion: why the poorest countries are failing and what can be done about it*. Oxford University Press.

Collins, H. and Evans, R. (2007) *Rethinking expertise*. University of Chicago Press.

Croll, J. (1875) *Climate and time*. Appleton & Co.: New York.

Crompton, T. (2008) *Weathercocks and signposts: the environment movement at a crossroads*. WWF Report: Godalming, UK.

Cronon, W. B. (1996) The trouble with wilderness: or, getting back to the wrong nature. In: Cronon, W. B. (ed.), *Uncommon ground: rethinking the human place in nature*. W.W. Norton & Co.: New York, pp. 69–90.

Cruikshank, J. (2005) *Do glaciers listen? Local knowledge, colonial encounters and social imagination*. UBC Press: Vancouver, Canada.

Daly, H. E. (1972) The economics of zero growth. *American Journal of Agricultural Economics* **54**(5), 945–54.

Dawson, T. P. and Allen, S. J. (2007) Poverty reduction must not exacerbate climate change. *Nature* **446**, 372.

Dessai, S., Adger, N. W., Hulme, M., Köhler, J., Turnpenny, J. and Warren, R. (2004) Defining and experiencing dangerous climate change. *Climatic Change* **64**, 11–25.

Diamond, J. (2005) *Collapse: how societies choose to fail or succeed*. Penguin: London.

Dietz, S., Hope, C. and Patmore, N. (2007) Some economics of 'dangerous' climate change: reflections on the Stern Review. *Global Environmental Change* **17**(3–4), 311–25.

Donner, S. (2007) Domain of the gods: an editorial essay. *Climatic Change* **85**, 231–6.

Douglas, M. and Wildavsky, A. (1982) *Risk and culture: an essay on the selection of technological and environmental dangers.* University of California Press: Berkeley, CA.

Doyle, J. (2007) Picturing the clima(c)tic: Greenpeace and the representational politics of climate change communication. *Science as Culture* **16**(2), 129–50.

Dunlap, T.R. (2004) *Faith in nature: environmentalism as religious quest.* University of Washington Press: Seattle, WA.

Dwivedi, O.P. (2001) Classical India. In: Jamieson, D. (ed.), *A companion to environmental philosophy.* Blackwell Publishers: Oxford, pp. 37–51.

Edwards, P.N. and Schneider, S.H. (2001) Self-governance and peer review in science-for-policy: the case of the IPCC Second Assessment Report. In: Miller, C.A. and Edwards, P.N. (eds), *Changing the atmosphere: expert knowledge and environmental governance.* MIT Press: Cambridge, MA, pp. 219–46.

Ereaut, G. and Segnit, N. (2006) *Warm words: how are we telling the climate story and can we tell it better?* Institute for Public Policy Research: London.

European Climate Forum (2003) *The biofuels directive: potential for climate protection? Conference Summary.* European Climate Forum: Tyndall Centre, UEA, Norwich, UK.

Evangelical Climate Initiative (2006) *Climate change: an evangelical call to action.* Evangelical Climate Initiative: Suwanee, GA.

Fagan, B. (2003) *The long summer: how climate changed civilisation.* Basic Books: New York.

Fankhauser, S., Tol, R.S.J. and Pearce, D.W. (1998) Extensions and alternatives to climate change impact valuation: on the critique of IPCC Working Group III's impact estimates. *Environment and Development Economics* **3**, 59–81.

Flannery, T. (2006) *The weather makers: our changing climate and what it means for life on Earth.* Penguin: London.

Fleming, J.R. (1998) *Historical perspectives on climate change.* Oxford University Press, Oxford and New York.

Fleming, J.R. (2006) The pathological history of weather and climate modification: three cycles of promise and hype. *Historical Studies in the Physical and Biological Sciences* **37**(1), 3–25.

Fleming, J.R. (2007) *The Callendar effect: the life and work of Guy Stewart Callendar (1898–1964).* American Meteorological Society: Boston, MA.

Ford, J.D. (2008) Where's the ice gone? Dimensions of vulnerability and adaptation in an Inuit community during anomalous sea-ice conditions. Paper presented at *Living with climate change: are there limits to adaptation?* Royal Geographical Society, London, 7–8 February 2008. Tyndall Centre: Norwich, UK.

Funtowicz, S. O. and Ravetz, J. R. (1993) Science for a post-normal age. *Futures* **25**, 739–55.

Furedi, F. (2007) *Invitation to terror: the expanding empire of the unknown.* Continuum Books: London.

Garvey, J. (2008) *The ethics of climate change: right and wrong in a warming world.* Continuum International Publishing: London and New York.

Glacken, C. (1967) *Traces on a Rhodian shore: nature and culture in Western thought from ancient times to the end of the eighteenth century.* University of California Press: Berkeley, CA.

Glacken, C. (1970) Man against nature: an outmoded concept. In: Helrich, H. W. (ed.), *The environmental crisis: man's struggle to live with himself.* Yale University Press: New Haven, CT.

Glover, L. (2006) *Postmodern climate change.* Routledge: London.

Golinski, J. (2003) Time, talk and the weather in eighteenth-century Britain. In: Strauss, S. and Orlove, B. J. (eds), *Weather, climate, culture.* Berg: Oxford, pp. 17–38.

Golinski, J. (2007) *British weather and the climate of enlightenment.* University of Chicago Press.

Goodall, A. H. (2008) Why have the leading journals in management (and other social sciences) failed to respond to climate change? *Journal of Management Inquiry* 17(4), 408–20.

Gore, A. (1992) On stabilising world population. *Population and Development Review* **18**(2), 380–3.

Greenpeace (1994) *Climate time bomb: signs of climate change from the Greenpeace database* Greenpeace, Amsterdam, Netherlands, 12pp.

Gribbin, J. (1977) *Forecasts, famines and freezes.* Walker: New York.

Grist, N. (2008) Positioning climate change in sustainable development discourse. *Journal of International Development* **20**(6), 783–803.

Grove, R. H. (1998) *Ecology, climate and empire: the Indian legacy in global environmental history, 1400–1940.* Oxford University Press: Delhi, India.

Haag, A. (2007) Climate change 2007: Al's army. *Nature* **446**, 723–4.

Haidt, J. (2006) *The happiness hypothesis.* Arrow Books: London.

Haidt, J. (2007) The new synthesis in moral psychology. *Science* **316**, 998–1002.

Halsnaes, K. and Shukla, P. (2007) Framing issues. Chapter 2 in *Climate change 2007: mitigation of climate change.* Report of Working Group III to the Intergovernmental Panel on Climate Change. Cambridge University Press.

Hann, J. (1908) *Handbuch der Klimatologie.* Stuttgart, Germany.

Hansen, J. E. (2007) Scientific reticence and sea level rise. *Environmental Research Letters* **2** doi:10.1088/1748–9326/2/2/024002.

Hawken, P. (2007) *Blessed unrest: how the largest movement in the world came into being and how no-one saw it coming.* Viking: New York.

Hillman, M. and Fox, C. (2007) Carbon rationing: a valuable way of cutting carbon emissions. *Science and Public Affairs* (June).

Hillman, M., Fawcett, T. and Raja, C. S. (2007) *The suicidal planet: how to prevent global climate catastrophe.* Thomas Dunne Books: New York.

Hobsbawm, E. (2007) *Globalisation, democracy and terrorism.* Little, Brown: London.

Holling, C. S. (1986) The resilience of terrestrial ecosystems: local surprise and global change. In: Clark, W. C. and Mann, R. E. (eds), *Sustainable development of the biosphere.* Cambridge University Press, pp. 217–32.

House of Lords (2000) *Science and society.* Report of the House of Lords Select Committee on Science and Technology: London.

Hulme, M. (2008) The conquering of climate: discourses of fear and their dissolution. *Geographical Journal* **174**(1), 5–16.

Hulme, M. (2009) A belief in climate. Chapter in Berry, S. (ed.), *Real scientists, real faith.* Lion Hudson: London.

Huntington, E. (2001) *Civilisation and climate* (reprinted from the 1915 edition). University Press of the Pacific: Honolulu, HI.

Huq, S. and Toulmin, C. (2006) *Three eras of climate change.* Sustainable Development Opinion Papers. International Institute for Environment and Development: London.

Imbrie, J. and Imbrie, K. P. (1979) *Ice ages: solving the mystery.* MacMillan: London.

Inglehart, R. (1997) *Modernisation and post-modernisation: cultural, economic and political change in 43 countries.* Princeton University Press: Princeton, NJ.

Inglehart, R. and Klingemann, H.-D. (2000) Genes, culture, democracy and happiness. In: Diener, E. and Suh, E. M. (eds), *Culture and subjective well-being.* MIT Press: Cambridge, MA, pp. 165–83.

IPCC (2000) *Emissions scenarios. Special Report of Working Group III of the Intergovernmental Panel on Climate Change.* Cambridge University Press.

IPCC (2007) *Climate change 2007: the physical science basis.* Contribution of Working Group I to the Fourth Assessment Report of the Intergovernmental Panel on Climate Change. Cambridge University Press.

James, O. (2007) *Affluenza.* Vermilion: Derry, NH.

Jamieson, D. (1996) Ethics and intentional climate change. *Climatic Change,* **33**, 323–36.

Jasanoff, S. (1990) *The fifth branch: science advisors as policy-makers.* Harvard University Press: Cambridge, MA.

Jasanoff, S. (2007) Technologies of humility. *Nature* **450**, 33.

Jenkins, K. (ed.) (1997) *The postmodern history reader.* Routledge: London.

Kellstedt, P. M., Zahran, S. and Vedlitz, A. (2008) Personal efficacy, the information environment, and attitudes towards global warming and climate change in the United States. *Risk Analysis* **28**(1), 113–26.

Kerr, R. (2007) Pushing the scary side of global warming. *Nature* **316**, 1412–15.

Kevane, M. and Gray, L. (2008) Darfur: rainfall and conflict. *Environmental Research Letters* 3 doi:10.1088/1748-9326/3/3/034006 10pp.

Kevles, D. (1985) *In the name of eugenics: genetics and the uses of human heredity.* Knopf: New York.

Killingsworth, M. J. and Palmer, J. S. (1996) Millennial ecology: the apocalyptic narrative from *Silent Spring* to global warming. In: Herndl, C. G. and Brown, S. C. (eds), *Green culture: environmental rhetoric in contemporary America.* University of Wisconsin Press: Madison, WI, pp. 21–45.

Kincer, J. B. (1933) Is our climate changing? A study of long-time temperature trends. *Monthly Weather Review* **61**(9), 251–9.

King, D. A. (2004) Climate change science: adapt, mitigate or ignore? *Science* **303**, 176–7.

Kuhn, T. (1962) *The structure of scientific revolutions.* University of Chicago Press.

Lamb, H. H. (1959) Our changing climate, past and present. *Weather* **14**, 299–318.

Lamb, H. H. (1977) *Climate: past, present and future. Vol. II: Climatic history and the future.* Methuen: London.

Lamb, H. H. (1982) *Climate, history and the modern world.* Methuen: London.

Langhelle, O. (2000) Why ecological modernisation and sustainable development should not be conflated. *Journal of Environmental Policy Planning* **2**, 303–22.

Latour, B. (2003) Is re-modernization occurring – and, if so, how to prove it? A commentary on Ulrich Beck. *Theory, Culture and Society* **20**(2), 35–48.

Latour, B. and Woolgar, S. (1979) *Laboratory life: the social construction of scientific facts.* Sage: Los Angeles, CA.

Leduc, T. B. (2007) Sila dialogues on climate change: Inuit wisdom for a cross-cultural interdisciplinarity. *Climatic Change* **85**, 237–50.

Leiserowitz, A. A. (2004) Before and after The Day After Tomorrow: a U.S. study of climate change risk perception. *Environment* **46**(9), 24–37.

Leiserowitz, A. A. (2006) Climate change risk perception and policy preferences: the role of affect, imagery and values. *Climatic Change* **77**, 45–72.

Lenton, T. M., Held, H., Krieglar, E., Hall, J. W., Lucht, W., Rahmstorf, S. and Schellnhuber, H.-J. (2008) Tipping elements in the Earth's climate system. *Proceedings of the National Academy of Sciences* **105**(6), 1786–93.

Lie, J. (2007) Global climate change and the politics of disaster. *Sustainability Science* **2**, 233–6.

Linder, S. H. (2006) Cashing-in on risk claims: on the for-profit inversion of signifiers for "global warming". *Social Semiotics* **16**(1), 103–32.

Livingstone, D. N. (2004) Climate. In: Thrift, N., Harrison, S. and Pile, S. (eds), *Patterned ground: entanglements of nature and culture.* Reaktion Books: London, pp. 77–9.

Lomborg, B. (ed.) (2004) *Global crises, global solutions.* Cambridge University Press.

Lomborg, B. (2007) *Cool it: the skeptical environmentalist's guide to global warming.* Cyan-Marshall Cavendish: London.

Lovelock, J. (2006) *The revenge of Gaia: why the earth is fighting back – and how we can still save humanity.* Penguin: London.

Lovelock, J. and Rapley, C. (2007) Ocean pipes could help the ocean to cure itself. *Nature* **449**, 403.

Lowe, T. D., Brown, K., Dessai, S., de Franca Doria, M., Haynes, K. and Vincent, K. (2006) Does tomorrow ever come? Disaster narrative and public perceptions of climate change. *Public Understanding of Science* **15**(4), 435–57.

Lowe, T. D. and Lorenzoni, I. (2007) Danger is all around: eliciting expert perceptions for managing climate change through a mental models approach. *Global Environmental Change* **17**(1), 131–46.

Lyotard, J.-F. (1984) *The postmodern condition: a report on knowledge.* University of Minnesota Press: Minneapolis, MN.

Malone, E. L. (2009) *From debate to agreement: the climate change issue.* Earthscan: London.

Malthus, T. (1798) *An essay on the principle of population, as it affects the future improvement of society, with remarks on the speculations of Mr. Godwin, M. Condorcet and other writers.* London.

Manabe, S. and Wetherald, R. T. (1975) The effects of doubling the CO_2 concentration on the climate of a general circulation model. *Journal of the Atmospheric Sciences* **32**, 3–15.

Martin, C. (2006) Experience of the New World and Aristotelian revisions of the Earth's climates during the Renaissance. *History of Meteorology* **3**, 1–16.

Martin, C. (2006) Artists on a mission. *Nature* **441**, 578.

McDermott, T. (2007) No time to lose. *New Statesman* (29 January).

McIntosh, A. (2008) *Hell and high water: climate change, hope and the human condition.* Birlinn: Edinburgh.

McKibbin, B. (1989) *The end of nature.* Random House: London.

Meadows, D. H. (1972) *The limits to growth: a report to the Club of Rome.* Universe Books: New York.

Meadows, D. H., Randers, J. and Meadows, D. L. (2004) *Limits to growth: the 30-year update.* Earthscan: London.

Merton, R. K. (1973) *The sociology of science: theoretical and empirical investigations.* University of Chicago Press.

Meyer, W. B. (2000) *Americans and their weather.* Oxford University Press.

Meyer, W. B. (2002) The Perfectionists and the weather: the Oneida Community's quest for meteorological utopia 1848–1879. *Environmental History* **7**(4), 589–610.

Michaelowa, A. and Michaelowa, K. (2007) Climate or development: is ODA diverted from its original purpose? *Climatic Change* **84**(1), 5–21.

Millais, C. (ed.) (2006) *Common belief: Australia's faith communities on climate change.* The Climate Institute: Sydney.

Miller, A. A. (1961) *Climatology* (9th edn). Methuen: London.

Miller, C. A. (2004) Climate science and the making of a global political order. In: Jasanoff, S. (ed.), *States of knowledge: the co-production of science and the social order.* Routledge: London, pp. 46–66.

Miller, C. A. and Edwards, P. N. (eds) (2001) *Changing the atmosphere: expert knowledge and environmental governance.* MIT Press: Cambridge, MA.

Millstone, E. (2005) Analysing the role of science in public policy-making. Chapter 2 in van Zwanenberg, P. and Millstone, E. (eds), *BSE: risk, science and governance.* Oxford University Press, pp. 11–38.

Mitford, N. (2006) *Love in a cold climate.* Penguin Classics: London. First published 1949.

Monbiot, G. (2006) *Heat: how to stop the planet burning.* Allen and Lane: London.

Monbiot, G., Lynas, M., Marshall, G., Juniper, T. and Tindale, S. (2005) Time to speak up for climate-change science. *Nature* **434**, 559.

Moran, D. D., Wackernagel, M., Kitzes, J. A., Goldfinger, S. H. and Boutaud, A. (2008) Measuring sustainable development: nation by nation. *Ecological Economics* **64**(3), 470–4.

Moser, S. (2007) More bad news: the risk of neglecting emotional responses to climate change information. In: Moser, S. and Dilling, L. (eds), *Creating a climate for change: communicating climate change and facilitating social change.* Cambridge University Press, pp. 64–81.

Moser, S. and Dilling, L. (eds) (2007) *Creating a climate for change: communicating climate change and facilitating social change.* Cambridge University Press.

Mudge, F. B. (1997) The development of the greenhouse theory of global climate change from Victorian times. *Weather* **52**, 13–17.

Munasinghe, M. and Swart, R. (2005) *Primer on climate change and sustainable development: facts, policy analysis and applications.* Cambridge University Press.

Murphy, B. F. and Timbal, B. (2008) A review of recent climate variability and climate change in southeastern Australia. *International Journal of Climatology* **28**(7), 859–80.

Murtaugh, P. (2009) Reproduction and the carbon legacy of individuals. *Global Environmental Change* (in press).

National Defense University (1978) *Climate change to the year 2000.* NDU: Washington DC.

National Development and Reform Commission (2007) *China's national climate change programme.* Beijing, China.

Neumayer, E. (2007) A missed opportunity: the Stern Review on climate change fails to tackle the issue of non-substitutable loss of natural capital. *Global Environmental Change* **17**(3–4), 297–301.

Newell, P. (2006) *Climate for change: non-state actors and the global politics of the greenhouse.* Cambridge University Press.

Nisbet, M. C. and Mooney, C. (2007) Framing science. *Science* **316**, 56.

Northcott, M. S. (2007) *A moral climate: the ethics of global warming.* Dartman, Longman and Todd: London.

Ogilvie, A. E. J. and Pálsson, G. (2003) Weather and climate in Icelandic sagas. In: Strauss, S. and Orlove, B. S. (eds), *Weather, climate, culture.* Berg: Oxford, pp. 251–74.

Okereke, C. and Bulkeley, H. (2007) *Conceptualizing climate change governance beyond the international regime: a review of four theoretical approaches.* Tyndall Centre Working Paper 112. University of East Anglia, Norwich, UK.

Olsen, K. H. (2007) The clean development mechanism's contribution to sustainable development: a review of the literature. *Climatic Change* **84**(1), 59–73.

O'Neill, B. C. and Oppenheimer, M. (2002) Dangerous climate impacts and the Kyoto Protocol. *Science* **296**, 1971–2.

O'Neill, B. C., MacKellar, F. L. and Lutz, W. (2001) *Population and climate change.* Cambridge University Press.

O'Neill, S., Osborn, T. J., Hulme, M., Lorenzoni, I. and Watkinson, A. R. (2008) Using expert knowledge to assess uncertainties in future polar bear populations under climate change. *The Journal of Applied Ecology* **45**(6), 1649–59.

Oppenheimer, M. (2005) Defining dangerous anthropogenic interference: the role of science, the limits of science. *Risk Analysis,* **26**(6), 1399–407.

O'Riordan, T. and Jordan, A. (1999) Institutions, climate change and cultural theory: towards a common analytical framework. *Global Environmental Change* **9**, 81–94.

Ott, H. E., Sterk, W. and Watanabe, R. (2008) The Bali roadmap: new horizons for global climate policy. *Climate Policy* **8**(1), 91–5.

Patt, A. (2007) Assessing model-based and conflict-based uncertainty. *Global Environmental Change* **17**(1), 37–46.

Pearce, D. (2003) The social cost of carbon and its policy implications. *Oxford Review of Economic Policy* **19**(3), 362–84.

Pearce, F. (2007) *With speed and violence: why scientists fear tipping points in climate change.* Beacon Press: Uckfield, UK.

Pettenger, M. (ed.) (2007) *The social construction of climate change.* Ashgate Press: Aldershot, UK.

Pettit, J. (2004) Climate justice: a new social movement for atmospheric rights. *IDS Bulletin* **35**(3), 102–6.

Philander, S. G. (2004) *Our affair with El Niño.* Princeton University Press: Princeton, NJ.

Pielke, R. Jr (2007) *The honest broker: making sense of science in policy and politics.* Cambridge University Press.

Polanyi, M. (1962) The republic of science: its political and economic theory. *Minerva* **1**, 54–74.

Ponte, L. (1976) *The cooling.* Prentice Hall: Englewood Cliffs, NJ.

Porritt, J. (2005) *Capitalism: as if the world matters.* Earthscan: London.

Porter, P. W. and Lukermann, F. E. (1976) The geography of utopia. In: Lowenthal, D. and Bowden, M. J. (eds), *Geographies of the mind: essays in historical geosophy in honour of John Kirtland Wright.* Oxford University Press, New York.

Posas, P. J. (2007) Roles of religion and ethics in addressing climate change. *Ethics in Science and Environmental Politics* **6**, 31–49.

Poumadere, M., Mays, C., Le Mer, S. and Blong, R. (2005) The 2003 heat wave in France: dangerous climate change here and now. *Risk Analysis* **25**(6), 1483–94.

President's Science Advisory Committee (1965) *Restoring the quality of our environment.* Report of the Environmental Pollution Panel (November): Washington DC.

Prins, G. and Rayner, S. (2007) *The wrong trousers: radically re-thinking climate policy.* Joint Discussion Paper of the James Martin Institute for Science and Civilisation, Oxford, and the MacKinder Centre for the Study of Long-Wave Events, London School of Economics October.

Prins, G. and Rayner, S. (2007) Time to ditch Kyoto. *Nature* **449**, 973–6.

Retallack, S. and Lawrence, T. (2007) *Positive energy: harnessing people power to prevent climate change.* Institute of Public Policy Research: London.

Reusswig, F. and Leiserowitz, A. (2005) The international impact of The Day After Tomorrow (Commentary). *Environment* **47**(3), 41–3.

Risbey, J.S. (2008) The new climate discourse: alarmist or alarming? *Global Environmental Change* **18**(1), 26–37.

Rittel, H. and Webber, M. (1973) Dilemmas in a general theory of planning. *Policy Sciences* **4(2)**, 155–69.

Roberts, J.T. and Parks, B.C. (2007) *A climate of injustice: global inequality, North–South politics and climate policy.* MIT Press: Cambridge, MA.

Robinson, J., Bizikova, L. and Cohen, S. (eds) (2008) Integrating climate change actions into local development. *Climate Policy* (special issue).

Robock, A. (2008) Twenty reasons why geo-engineering may be a bad idea. *Bulletin of the Atomic Scientists* **64**(2) (May/June), 14–18, 59.

Roe, G.H. and Baker, M.B. (2007) Why is climate sensitivity so unpredictable? *Science* **318**, 629–32.

Ross, A. (1991) Is global culture warming up? *Social Text* **28**, 3–30.

Sachs, J.D. (2004) Seeking a global solution. *Nature* **430**, 725–6.

Sachs, J.D. (2005) *The end of poverty: how we can make it happen in our time.* Penguin Books: London.

Sanders, T. (2003) (En)Gendering the weather: rainmaking and reproduction in Tanzania. Chapter 5 in Strauss, S. and Orlove, B. (eds), *Weather climate culture.* Berg: Oxford, pp. 83–102.

Sanderson, M. (1999) The classification of climates from Pythagoras to Köppen. *Bulletin of the American Meteorological Society* **80**(4), 669–73.

Sarewitz, D. and Pielke, R.J. Jr (2006) The neglected heart of science policy: reconciling supply of and demand for science. *Environmental Science and Policy* **10**, 5–16.

Schellnhuber, H.-J., Crutzen, P., Clark, W.C., Claussen, M. and Held, H. (eds) (2004) *Earth system analysis for sustainability.* MIT Press: Cambridge, MA.

Schellnhuber, H.-J., Cramer, W., Nakicenovic, N., Wigley, T.M.L. and Yohe, G. (eds) (2006) *Avoiding dangerous climate change.* Cambridge University Press.

Schneider, S.H. (1985) Science by consensus: the case of the NDU study 'climate change to the year 2000' – an editorial. *Climatic Change* **7**, 153–7.

Schneider, S. H. (1989) *Global warming: are we entering the greenhouse century?* Sierra Club Books: San Francisco, CA.

Schneider, S. H. (1998) Kyoto Protocol: the unfinished agenda. *Climatic Change* **39**, 1–21.

Schneider, S. H. and Londer, R. (1984) *Coevolution of climate and life.* Sierra Club Books: San Francisco, CA.

Schneider, S. H. and Mesirow, L. E. (1976) *The genesis strategy: climate and global survival.* Dell: New York.

Schneider, S. H., Semenov, S. and Patwardhan, D. (2007) Assessing key vulnerabilities and the risk from climate change. In: Parry, M. L., Canziani, O. F., Palutikof, J. P., van der Linden, P. J. and Hanson, C. E. (eds), *Climate change 2007: impacts, adaptation and vulnerability. Contribution of Working Group II to the Fourth Assessment Report of the Intergovernmental Panel on Climate Change.* Cambridge University Press, pp. 779–810.

Schwartz, P. and Randall, D. (2003) *An abrupt climate change scenario and its implications for US national security.* Report by the Global Business Network prepared for the US Department of Defense, October.

Schwartz, S. H., Melech, G., Lehmann, A., Burgess, S., Harris, M. and Owens, V. (2001) Extending the cross-cultural validity of the theory of basic human values with a different method of measurement. *Journal of Cross-Cultural Psychology* **32**(5), 519–42.

Scott, A. (2007) Peer review and the relevance of science. *Futures* **39**, 827–45.

Segnit, N. and Ereaut, G. (2007) *Warm words II: how the climate story is evolving.* Institute for Public Policy Research/Energy Savings Trust: London.

Shanahan, M. (2007) *Talking about a revolution: climate change and the media. COP13 Briefing and Opinion Papers.* IIED: London.

Shearman, D. and Smith, J. W. (2007) *The climate change challenge and the failure of democracy.* Greenwood Press: Westport, CT.

Simms, A. and Smith, J. (eds) (2008) *Do good lives have to cost the earth?* Constable: London.

Singer, F. C. and Avery, D. T. (2007) *Unstoppable climate change every 1500 years.* Rowman and Littlefield: Lanham, MD.

Slocum, R. (2004) Polar bears and energy-efficient light-bulbs: strategies to bring climate change home. *Environment and Planning D: Society and Space* **22**, 413–38.

SMIC (1971) *Inadvertent climate modification. Report of the Study of Man's Impact on Climate (SMIC).* MIT Press: Cambridge, MA.

Smith, D. M. (2006) *Just one planet: taking action on poverty and climate change.* Practical Action: Rugby, UK.

Smith, J. (ed.) (2000) *The daily globe: environmental change, the public and the media.* Earthscan: London.

Smith, J. (2005) Dangerous news: media decision making about climate change risk. *Risk Analysis* **25**, 1471–82.

Sneddon, C., Howarth, R. B. and Norgaard, R. B. (2006) Sustainable development in a post-Brundtland world. *Ecological Economics* 57(2), 253–68.

Spash, C. L. (2005) *Greenhouse economics: values and ethics*. Routledge: London.

Spash, C. L. (2007) The economics of climate change impacts à la Stern: novel and nuanced or rhetorically restricted? *Ecological Economics* 63(4), 706–13.

Stern Review (2006) *The economics of climate change*. Cambridge University Press.

Strauss, S. and Orlove, B. (eds) (2003) *Weather, climate, culture*. Berg: Oxford.

Taylor, B. (ed.) (2005) *Encyclopedia of religion and nature*. Thoemmes Continuum: London and New York.

Tennekes, H. (1990) A sideways look at climate research. *Weather* 45, 67–8.

Thompson, M. and Rayner, S. (1998) Cultural discourses. In: Rayner, S. and Malone, E. L. (eds), *Human choice and climate change. Vol. 1: The societal framework*. Battelle Press: Columbus, OH, pp. 265–344.

Thompson, M. and Rayner, S. (1998) Risk and governance. Part I: The discourses of climate change. *Government and Opposition* 33(2), 139–66.

Thompson, M., Rayner, S. and Ney, S. (1998) Risk and governance. Part II: Policy in a complex and plurally perceived world. *Government and Opposition* 33(2), 330–54.

Toman, M. A. (2005) Climate change mitigation: passing through the eye of the needle? In: Sinnott-Armstrong, W. and Howarth, R. B. (eds), *Perspectives on climate change: science, economics, politics and ethics*. Elsevier: Amsterdam, pp. 75–98.

Toman, M. A. (2006) Values in the economics of climate change. *Environmental Values* 15, 365–79.

Toynbee, A. J. (1934) *A study of history. Vol. II: The genesis of civilizations*. Oxford University Press.

Trumbo, C. W. and Shanahan, J. (2000) Social research on climate change: where we have been, where we are, and where we might go. *Public Understanding of Science* 9, 199–204.

Tyndall, J. (1855–1872) *Journals of John Tyndall Vol. 3, 1855–1872*. Unpublished Tyndall Collection, Royal Institution, London.

Tyndall, J. (1861) On the absorption and radiation of heat by gases and vapours. *Philosopical Magazine* (Ser. 4) 22, 169–94 & 273–85.

Tyndall, J. (1863) On the passage of radiant heat through dry and humid air. *Philosopical Magazine* (Ser. 4) 26(172), 44–54.

US Conference of Catholic Bishops (2001) *Global climate change: a plea for dialogue, prudence and the common good*. United States Conference of Catholic Bishops: Washington DC.

Velders, G. J. M., Anderson, S. O., Daniel, J. S., Fahey, D. W. and McFarland, M. (2007) The importance of the Montreal Protocol in protecting climate. *Proceedings of the National Academy of Sciences* 104(12), 4814–19.

Verweij, M. and Thompson, M. (eds) (2006) *Clumsy solutions for a complex world: governance, politics and plural perspectives*. Palgrave Macmillan: Basingstoke, UK, and New York.

Victor, D.G. (2001) *The collapse of the Kyoto Protocol.* Princeton University Press: Princeton, NJ, and Oxford.

Victor, D.G., House, J.C. and Joy, S. (2005) A Madisonian approach to climate policy. *Science* 309, 1820–1.

von Neumann, J. (1955) Can we survive technology? *Fortune (June)*. Reproduced in *Population and Development Review* (1986) 12(1), 117–26.

Von Storch, H. and Stehr, N. (2006) Anthropogenic climate change: a reason for concern since the 18th century and earlier. *Geografiska Annaler* 88A(2), 107–13.

Vonk, M.A. (2006) *The quest for sustainable lifestyles and quality of life: contributions from Amish, Hutterite, Franciscan and Benedictine philosophy of life.* Paper presented at the conference *Exploring religion, nature and culture*, Gainesville, FL, L. 6–9 April.

Walker, G. and King, D. (2008) *The hot topic: how to tackle global warming and still keep the lights on.* Bloomsbury Press: London.

Weart, S.R. (1997) Global warming, cold war and the evolution of research plans. *Historical Studies in Physical Biological Sciences* 27(2), 319–56.

Weart, S.R. (2003) *The discovery of global warming.* Harvard University Press: Cambridge, MA.

Weart, S.R. (2007) Money for Keeling: monitoring CO_2 levels. *Historical Studies in the Physical and Biological Sciences* 37(2), 435–52.

Weber, E.U. (2006) Experience-based and description-based perceptions of long-term risk: why global warming does not scare us (yet). *Climatic Change* 77, 103–20.

Weber, M. (1946) *Essays in sociology.* Gerth, H.H. and Mills, C.W. (trans/eds). Oxford University Press.

Weiss, H. and Bradley, R.S. (2001) What drives societal collapse? *Science* 291, 610.

White, L. Jr (1967) The historical roots of our ecological crisis. *Science* 155, 1203–7.

Whitmarsh, L. (2008) What's in a name? Commonalities and differences in public understanding of "climate change" and "global warming". *Public Understanding of Science* doi:10.1177/0963662506073088.

World Commission on Environment and Development (1987) *Our common future.* Oxford University Press.

WWF (World Wide Fund for Nature) (2008) *2010 and beyond: rising to the biodiversity challenge.* WWF: Gland, Switzerland.

Yearley, S. (2006) How many "ends" of nature: making sociological and phenomenological sense of the end of nature. *Nature and Culture* 1(1), 10–21.

Index